THE INFINITE IN GIORDANO BRUNO

THE INFINITE IN
Giordano Bruno

WITH A TRANSLATION OF HIS DIALOGUE

Concerning the Cause
Principle, and One

BY SIDNEY GREENBERG

King's Crown Press
COLUMBIA UNIVERSITY, NEW YORK
1950

KING'S CROWN PRESS is a subsidiary imprint of Columbia University Press established for the purpose of making certain scholarly material available at minimum cost. Toward that end, the publishers have adopted every reasonable economy except such as would interfere with a legible format. The work is presented substantially as submitted by the author, without the usual editorial and typographical attention of Columbia University Press.

PUBLISHED IN GREAT BRITAIN, CANADA, AND INDIA
BY GEOFFREY CUMBERLEGE, OXFORD UNIVERSITY PRESS
LONDON, TORONTO, AND BOMBAY

MANUFACTURED IN THE UNITED STATES OF AMERICA

To My Father and Mother

PREFACE

THIS BOOK is not intended to be a critique of the philosophy of Giordano Bruno. Its purposes are: first, to give a faithful account of Bruno's thought as expressed in some of his writings; second, to give an analysis and exposition of the thought of Bruno, as he himself developed it, with regard to the problem of the infinite; third, to give the first English translation of the "De la causa," thus allowing the reader to turn to the text itself.

The author intends to supplement this work with a critique of the philosophy of Giordano Bruno, with special emphasis on its contributions to the Renaissance philosophy of science.

I should like to express my thinks and appreciation to Professors P. O. Kristeller, J. H. Randall, Jr., Dino Bigongiari, and Ernest A. Moody, of Columbia University, for their advice and guidance in the preparation of this work. Special thanks are due to Professor Kristeller for his generous assistance in the task of correcting and revising the translation.

Acknowledgement is gratefully made to Charles Scribner's Sons for permission to quote from E. Gilson, *The Spirit of Medieval Philosophy*, and to Charles T. Branford Company for permission to quote from Stephen McKenna's translation of Plotinus.

S. G.

Brooklyn, N.Y.
January, 1950

CONTENTS

THE INFINITE IN GIORDANO BRUNO

INTRODUCTION

An acquaintance with a thing's incidental properties greatly promotes an understanding of its essential nature; we are best qualified to speak of a thing's nature when we are able to give an account of its properties as they are experienced. It was Aristotle who said that we come to the knowledge of the nature of a thing through a study of its operations.[1] The order of knowledge is such that it necessarily proceeds from operations to that which lies beneath. Though we arrive at the substance last in the order of knowledge, the substance is first in the order of existence. Hence, the more we know of these operations, the more we can know of the substance. Knowledge of a thing may thus increase and may be expected to increase as long as (1) the knower does not fall into the error of identifying the operations with the thing itself and (2) the knower does not emphasize one operation to the exclusion of all the others.

We do not believe it is otherwise with the knowledge gained through the study of a man and his philosophy, for this also proceeds by way of analysis and the study of the many approaches to the man. We believe, indeed, that in an investigation of this kind the tendency is too often to exclude another's approach, while substituting one's own. It is a case of "either-or," when it is to greater advantage to be "both-and." For this reason, we believe that in pursuing such an analysis, there are two pitfalls which we must avoid: (1) we must not hesitate to accept, as a basis from which to work, all the previous ways of approaching the subject, with the firm conviction that each contributes to the knowledge of that subject; and (2) we must not identify these ways, nor our own way, with the subject.

Our first task, therefore, is to trace the various methods employed by previous interpreters; by so doing, we hope to reach that point where yet another way may be added. These methods will be treated in terms of two periods. The first period is that of the restoration of Bruno's philosophy to its rightful place in the history of thought, a period which is concerned with establishing Bruno's existence, and with reestablishing his reputation. This is a period which employs the instrument of documentation, and which extends its analysis to Bruno's influence on the history of philosophy. Representative of this period are, among others, Bartholmess,[2] Carrière,[3] Berti,[4] Clemens,[5] Brunnhofer,[6] and Wagner.[7]

The second period is that which is marked by the detailed examination of Bruno's works, and with further documentation as means of aiding the interpretation of his philosophy. Included in this effort are the first real attempts to treat the basic problems and the principles of Bruno's philosophy, with a view to establishing some systematization. As representative of this second period of Bruno's studies, we may mention among others, Tocco,[8] Gentile,[9] Mondolfo,[10] Troilo,[11] Olschki,[12] Spampanato,[13] Salvestrini,[14] Namer,[15] Sarauw,[16] De Ruggiero,[17] Corsano,[18] Guzzo,[19] Cassirer,[20] McIntyre,[21] Frith,[22] and the publication of the Latin works.[23]

The first definitive steps to further an understanding of Giordano Bruno and his philosophy were not undertaken until the late eighteenth century and early nineteenth century. This was due to the fact that Bruno's philosophy had been vilified, and his reputation damned, by the few studies made in the one hundred years prior to 1785. C. Bartholmess, in his two-volume work on Bruno, has traced the sequence of events which led to the necessity of a restoration. Pierre Bayle's article, in his *Dictionnaire historique et critique* (1697), erected a wall around Bruno's philosophy which defied penetration, mainly because it was Bayle's thesis that both Bruno and Spinoza were exponents of "the damnable doctrine of atheism."[24] That Bayle had no real acquaintance with Bruno's work and with Bruno's thought is evident from the fact that his summaries are inexact; the result was that the account given by Bayle "was twisted out of proportion."[25] In England too, Bruno's reputation was subjected to slanderous attacks, and though John Toland tried to salvage something of the spirit of Bruno,[26] the impact of these opponents was more than he could overcome. Until the advent of Jacobi's "Letter on Spinoza's Philosophy," in 1785, the situation remained static; after Jacobi, whose work marks the restoration of Bruno's philosophy, there came welcome respect from Goethe,[27] and from such popular historians of philosophy as Buhle and Tennemann. It was the opinion of these historians that Bruno "achieved a philosophy of the Absolute, two centuries before Schelling and Hegel." Schelling himself regarded Bruno as his forerunner,[28] while Hegel,[29] though he did not share the enthusiasm of Schelling, became the direct source of the first real studies on Bruno's philosophy. The first of these was that of Bartholmess, and the second that of M. Carrière. "With few materials at their command, the two young scholars overcame the contention, raised by the Catholics, that Bruno never existed, or if he existed, was never burnt."[30] For Bartholmess, Bruno was "the disciple of Pythagoras, the champion of Plato, Plotinus, and Proclus, the successor to Raymond Lull, the apologist for Copernicus, the admirer of Tycho Brahe, and the precursor of Spinoza."[31] To the method of investigation and analysis employed by Bartholmess, Carrière added the method of philosophical interpretation which laid emphasis, first, on the importance for the history of philosophy of Bruno's conception of the transcendence and immanence of God, and secondly, on the influence of Bruno's philosophy on Descartes, Leibniz, Kant, and Hegel. For Carrière, Bruno is above Spinoza in his teaching of the infinity of substance, and above Leibniz in his teaching of the theory of monads.[32]

The studies of Bartholmess and Carrière not only served to restore Bruno's reputation, but also set the stage for further investigation. Their studies, in conjunction with the documentary research of Berti, gave impetus to the more detailed work of Brunnhofer, whose treatment not only continued and enlarged upon that of Bartholmess, Carrière, and Berti, but also showed that the influence of Bruno extended far beyond his immediate environs.[33] Though Brunnhofer has been accused of allowing his own feelings to veil the facts,[34] his attitude can best be understood in terms of his own statement: "Though the investigators of today may smile at Bruno's theories, they cannot discount the enduring charm in his poems, the concept of the unity of the universe, the belief in a life on all the planets, and the concept of the connection between the processes of mind and body."[35]

With the arrival of more detailed studies came the judgment that Bruno's philosophy defied systematic treatment; the problem of inconsistency and inner contradiction thus became the keynote of the second period of Bruno's scholarship. Tocco undertook the task of pointing out the inconsistencies, and sought to explain them as the result of Bruno's intellectual development.[36] The development theory formulated by Tocco gave impetus to a lively scholarly discussion. The best of Bruno's English interpreters considered Bruno's philosophy to be a conglomerate mosaic of heterogeneous ideas, having within themselves obscurities of doctrine and language.[37] Olschki finds that Bruno attached no great importance to terminology, so that there is no fixed meaning in the philosophy; in fact, says Olschki, Bruno glories in complexities, aphorisms, long sentences, and enumerations, to such a degree that they becloud the issue and make it difficult to affirm or deny his argument.[38] In Mrs. Frith's opinion, too many of Bruno's interpreters bestow their own individuality upon his philosophy, so that Bruno will be found to appear as a materialist, an atheist, a pantheist, and sometimes in his true character as an idealist.[39]

We may now undertake the task of surveying the various resolutions that have been offered to the problem of inconsistency and contradiction.

In this task special emphasis is to be placed upon the work of Tocco, because he has been the center of attention in the controversy.

Tocco declares that the nine years (1582-1591) between the publication of the "De umbris idearum" and the "De minimo" Bruno passed from the philosophy of Plotinus to the philosophy of Democritus. Between these two extremes, and as their point of union, according to the author of the development theory, stands the pantheism of the dialogue "De la causa," and of the other Italian works produced in the years 1584-1585.[40] The "De umbris idearum" has, as its principal features, Neoplatonic emanationism, universal animism, and the doctrine of the transcendence of the One.[41] In this same period falls the "Sigillus sigillorum," with its teaching of the theory of "Ideas" as the transcendent source of sensible things.[42] As one work leads to another, and as one period leads to another, says Tocco, the strain becomes more evident; though it cuts itself loose from some emphases, as it proceeds, the strain remains a continuous one. Hence, the traces of Neoplatonism remain even in the "De la causa," though the latter is the representative of the middle period of Bruno's "development."[43] But in the "De la causa" these traces are less prominent and are subordinated to new tendencies which are monistic and pantheistic. Here the "Ideas" are no longer transcendent[44]; they are the formal causes of things, the directive norms which the "Universal Intellect" employs in its operations. Here, too, form and matter are held to be substantially one; they are two aspects of the same Substance[45]; while for Neoplatonism the One is inconceivable because it is above nature, Bruno is said to have held in the "De la causa" that the One is inconceivable because it is confused with nature.[46] From Neoplatonism, which considers nature as the shadow of reality, Bruno thus returns to the pre-Socratic monism, which sees in nature the true reality. It is for this reason, says Tocco, that there is in the "De la causa" an attempted reconciliation between Parmenides, with his doctrine of an unchangeable substance, and Heraclitus, with his teaching of the ever-changing flux.[47] In the "De minimo," which Tocco takes as representative of the last period of Bruno's development, there are continuing strains of Neoplatonism, although not as strong as in the "De la causa." According to Tocco, although the "De minimo" contains a supreme triad of God, Nature, and Intellect, as against the One, Mind, and Soul, of Plotinus, Bruno now stresses the fact that "Nature" is still to be identified with God, just as in the "De la causa." While the "De minimo" is thus closer to the "De la causa" and to Parmenides and Heraclitus, than to the "De umbris" and to Plotinus, yet it is different from the "De la causa" in that it adds another factor, the concept of the atom.[48]

In order to preserve, reconcile, and develop this conception, Bruno chose the name "monad" to emphasize the change, but indicated that the atom as he conceived it is neither the Unity of the Pythagoreans, nor the infinitely small particle of Democritus (though it has an affinity with it); it is, at one and the same time, soul and body, material subject, and center of energy. The "strain" continues throughout, for to his new concept of the atom, Bruno has attached the older concept found in "De la causa" of a World Soul resting in itself, and yet diffusing itself throughout every atom.[49] Tocco maintains that Bruno is not conscious of the historical divergences that have been admitted; though he may admit his eclecticism, his plan is to rid the doctrines of contradiction by holding to what is common—and thus to advance.[50]

Against the position outlined by Tocco, and shared essentially by Mondolfo and Troilo, there is the side upheld by Gentile, and accepted in principle by Olschki and McIntyre. These scholars declare themselves against any positive attribution of development to Bruno's philosophy, by pointing out that there can be no unity in a system which is characterized by oscillations and contradictions. Where, they ask, in a philosophy that glories in complexities, aphorisms, enumerations and analogies, gleaned through eclecticism and syncretism, is the unity that Tocco affirms?

For Gentile, there is no unity, coherence, or development in a philosophy that affirms, at one and the same time, the doctrines of Parmenides and Heraclitus, the combination of Eleatic monism and Neoplatonism, the doctrine of Universal Soul and the mechanistic theory of Democritus, the teachings of Spinoza regarding the concept of substance, the Leibnizian monadology, and both the positive and negative positions in religion. These declarations are more fitting to the thought of Tocco than to the philosophy of Bruno.[51] An example of this, says Gentile, is the simultaneous affirmation of positive and negative religion; the "contradiction" employed to show Bruno's development from one to the other is one that is superimposed on the philosophy of Bruno. For Gentile, the principle that represents Bruno's philosophy in its fullest sense is "mens insita rebus"; no one of the pages of Bruno can be understood without this principle. But with this principle, the contradiction attributed to Bruno is removed, and with it also the "development." In Bruno's philosophy, there is not one stage in which he declares himself for positive religion and another stage in which he advocates the negative position. The theism of Bruno is not "his religion, but the limit of his philosophy; Bruno's true religion is essentially pantheistic." Bruno had always held, says Gentile, that philosophy and reason alone determine the truth: "The contemplation which rests on belief is nothing."[52]

Olschki not only disagrees with Tocco, but also with Gentile; for it is his belief that Bruno's philosophy is so confused that it is impossible to gain, by any method, any fixity of meaning. Bruno's habit of enumeration beclouds every issue, whether it deals with physics, metaphysics, or morals.[53] In spite of the fact that since Carrière there have been new additions to and new publications of the works of Bruno, this major criticism still stands: that Bruno's thought is not susceptible of a consistent interpretation.[54] Bruno's oscillations between divergent and even contradictory positions makes it impossible, in spite of Tocco, to attribute to him a definite doctrine: "Bruno is fantastic, arbitrary, and unintelligible; he has too many points of view, to have one true one."

The writer considers it his task to supplement and correct these findings with a faithful account of Bruno's thought, as expressed in some of his writings. Two methods will be employed in this task. First, we shall give an exposition and analysis of the thought of Bruno, as he himself developed it, with regard to a major problem. For this we have chosen the "De la causa" and the "De l'infinito" because they concern themselves with the problem of the Infinite. This will call for a more detailed interpretation of these two works than has ever been given in English. Secondly, we shall give the first English translation of the "De la causa," thus permitting the reader to turn to the text itself. We believe that this is not only the best procedure, but also that it is a necessary one; we believe that a "commentary" should be accompanied by that of which it is a commentary, since otherwise the commentary becomes a far-removed exposition bordering on the status of a report.

We said above that none of the ways of approaching the meaning of Bruno's philosophy must be sacrificed to the exclusion of another. We therefore accept all of these ways, because we believe that Bruno's philosophy is representative of all and that Bruno purposely pursued the paradoxical in order to formulate his doctrine. Ours is a method which presupposes a unity in principle, regardless of apparent inconsistencies in detail.

That the problem of the Infinite is a fundamental one for Bruno is evident from the fact that its resolution is the goal of at least five of his major works. Two of these, the "De la causa" and the "De l'infinito," are the subject matter of this thesis; a third, the "De immenso," is the Latin equivalent of the "De l'infinito"[55]; a fourth, "La cena de le ceneri," treats the Copernican theory from the point of view of its relationships to the new metaphysics[56]; a fifth, the "Eroici furori," has as its major concern the moral significance of the concept of Infinity. The "Eroici furori" is especially important for us, for it reveals Bruno's answer to the question: Why study the problem of Infinity at all? It will serve us in two ways: first, as a background against which to study the problem; and, secondly, as a means of introducing the problem itself, in terms of its objectives.

It is in the contemplation of the Infinite, that man attains his greatest good. Since all things strive toward the end which is intended for them by nature, the more perfect the nature, the more perfect is the tendency to fulfillment; since human tendency and aspiration cannot find its fulfillment in finite goods, and in finite truths, it is a fact that the human intellect and the human will are never at rest. The final goal, therefore, is not to be found in particular goods, for these lead the individual from one thing to another, and merely make more evident the fact that there is more good to be desired, and more truth to be known. In each person, there is an innate desire to become all things; for this reason, man is directed by will and intellect to the Infinite, which is at once his cause, source, and end. The existence of this infinite desire, and of this infinite quest for knowledge, implies the existence of that which can satisfy both.[57] There waits for each being, says Bruno, eternity and realization[58]; the contemplation of the universe is the means whereby the person rises to the contemplation of the true Infinite Being.[59] The goal and end of the person is to become united with its eternal source, to escape from relativity and change, to truth, eternity, and immutability[60]; this aim is achieved by an ascent from darkness to the light of the sun, the Infinite Light[61]; for this act a conversion is necessary in order to expel the lower feelings of sense which deal only with the particular.

It is the presence of the Infinite in man that compels him to love the Infinite, and thereby to become one with it.[62] The true man must conform himself to the pattern of the Infinite, diverting his sight from the things which stand between himself and his perfection. He must apply himself with full intention to superior things. He must bring his entire will and all his affections under the influence of the Infinite, which is the final and perfect object of love.[63] The intellectual power is satisfied with comprehended truth only by advancing ever nearer and nearer to incomprehensible truth.[64] This search for the Infinite is from the finite or measured to the illimitable and the immeasurable; from the "contracted this" to the Infinite Substance; from the good, true, and beautiful—by participation—to the one true, good, and beautiful Being itself.

The object of the intellect and the will is a positive or a perfective Infinite, which means that the intellect and the will of man can perfect themselves by means of this "object." Without the proper object, man pursues the privative infinite of potentiality, which is the fleeting ghost of the particular.[65] In other words, the will does not find rest in a finite good, nor does the intellect find rest in a finite truth; the will therefore seeks that which is the source of the present good, and the intellect seeks that which is the source of the present truth. Human perfection consists in a form of knowledge by which the knower becomes one with the Infinite: "Acteon came into the presence of the same, and ravished out of himself by so much splendor, he became the prey, saw himself converted into that which he was seeking, and perceived that of his dogs or thoughts, he himself became the longed for prey; for, having absorbed the infinite divinity into himself, it was no longer necessary to search outside himself for it."[66] "To see the infinite is to be seen by the infinite."[67]

What, then, is the true calling of man? This much is evident from the "Eroici furori": man's perfection lies in seizing the supreme truth by means of reason, and in practising sovereign good by means of the will. Because man cannot endure that which is divided, fleeting, or in part imperfect, he looks for all things to be full, lasting, universal, and necessary; the need of infinite perfection is no caprice or superfluity of thought, but the real and lasting—the most noble and the most lawful of all our desires. The whole creation, in all its splendor, offers us satisfaction; and since it is the high calling of man to comprehend the universe, "let him raise his eyes and his thoughts to the heavens which surround him, and the flying worlds above; they are a picture, a book, a mirror, wherein he can behold and read the forms and the laws of the supreme goodness, the plan and the total of perfection."[68]

The "De la causa" and the "De l'infinito" are the products of Bruno's attempt to "comprehend the universe"; they are the results of Bruno's desire to fulfill that "high calling of man" which is the contemplation of the Infinite—for "it is in the contemplation of the Infinite that man attains his greatest good."

In Bruno's exposition of his conception of the Infinite, we shall find elements which will be recognized as those of his predecessors, with some of whom he agrees in part, like Cusanus and Lucretius; with others of whom he largely disagrees, like Aristotle. These serve only as a background for Bruno's view, a view which tries not only to analyze and synthesize, but also to add, in such a way that Bruno's own view reflects all of them. Bruno is the artist who, in painting the picture of the Infinite, employs brushes which are sometimes Aristotelian, sometimes Cusanian, sometimes Plotinian; yet the picture resembles Bruno rather than the brushes, which remain but instruments in the hands of one who has them at his command.

The consistent application of our principle, however, does not allow us to go on without surveying the field again; that is, since an acquaintance with a thing's incidental properties greatly promotes an understanding of its essential nature, we must now turn our attention to the historical background of the problem of the Infinite.

The problem of the infinite, says Aristotle, is one that the natural philosopher cannot ignore. Although, in general, not everything can be called infinite or finite, there are many considerations that demand some sort of an infinite as an explanation; among these are: (1) time, which is infinite; (2) the division of magnitude; (3) if coming to be and passing away are not to give out, the source of coming to be must be infinite; (4) the fact of limitation, since everything finite is limited by something which is in turn limited by something else; (5) not only number, but also mathematical magnitudes, and what is outside the heaven are supposed to be infinite, because they never give out in our thought. All these facts serve to persuade us that the infinite in some way exists, and previous investigation also emphasizes the importance of the problem. Proceeding along the lines of his other investigations, Aristotle seeks to derive from his predecessors all of the possible formulations of this problem.[69]

As a point of departure, he lists the different senses in which the term infinite is used: (1) what is incapable of being gone through, because it is not its nature to be gone through; (2) what admits of being gone through, the process, however, having no termination; (3) what scarcely admits of being gone through; (4) what naturally admits of being gone through, but is not actually gone through or does not actually reach an end. Further, everything that is infinite may be so in respect to addition or division or both.[70]

After distinguishing these senses in which the term infinite is used, Aristotle examines the earlier formulation of the problem. First, there is that resolution which declares the infinite to be a principle in its own right.

> Some, as the Pythagoreans and Plato, make the infinite a principle in the sense of a self-subsistent substance, and not as a mere attribute of some other thing. Only the Pythagoreans place the infinite among the objects of sense (they do not regard number as separable from these) and assert that what is outside the heaven is infinite. Plato, on the other hand, holds that there is no body outside (the Forms are not outside since they are nowhere) yet that the infinite is present, not only in the objects of sense, but in the Forms also.[71]

Aristotle holds that the infinite refers to quantity and to refer it to substance is an abandonment of the facts and leads to contradictions.

> It is plain, too, that the infinite cannot be an actual thing and a substance and principle. For any part of it that is taken will be infinite, if it has parts: for to be infinite and the infinite are the same, if it is substance and not predicated of a subject. Hence, it will be either indivisible or divisible into infinites. But the same thing cannot be many infinites. (Yet just as part of air is air, so a part of the infinite would be infinite, if it is supposed to be a substance and principle.) Therefore, the infinite must be without parts and indivisible. But this cannot be true of what is infinite in full completion; for it must be a definite quantity.[72]

"Thus the view of those who speak after the manner of the Pythagoreans is absurd. With the same breath they treat the infinite as substance, and divide it into parts."[73]

For the Pythagoreans, number comprises within itself two species, the odd and the even; it is, therefore, the unity of these two contraries; it is the odd and the even. Now, one, or the unit, is the odd and the even; hence, the one, or the unit, is the essence of number. Further, since unity and infinity can be predicated of all, all is one and infinite. But one and infinite, odd and even, is number; therefore, all is number.[74]

"But Plato has two infinites, the Great and the Small."[75] Believing the "Ideas" to be the causes of everything, Plato thought that their constituent elements were the elements of everything; their material principles were the Great and the Small, but their formal principle was the One. "As matter the great and the small were principles; as essential reality, the One; for from the great and the small, by participation in the One come the numbers."[76] In regarding the One as substance, and not as a predicate of something else, Plato's doctrine resembled that of the Pythagoreans, as it did also in treating the numbers as the causes of being in everything else.[77] But, says Aristotle, it was peculiar to Plato to set up a duality instead of a simple unlimited, and to make the Unlimited consist in the Great and the Small. Peculiar to Plato, too, was the doctrine that the numbers are distinct from sensible things, whereas, for the Pythagoreans, things are numbers.

> But positing a dyad and constructing the infinite out of great and small, instead of treating the infinite as one, is peculiar to him; and so his view that the numbers exist apart from sensible things, while they say that the things themselves are numbers, and do not place the objects of mathematics between Forms and sensible things.[78]

Aristotle's arguments against Plato, like those against the Pythagoreans, are directed in this context against the conception of the infinite as an actual substance.[79] "Now it is impossible that the infinite should be a thing which is itself infinite, separable from sensible objects."[80]

Aristotle's arguments against both Plato and the Pythagoreans have taken the form of reducing the conception of an actual infinite to absurdity. This "resolution" has yielded only negative results. We shall have to turn our attention to that formulation which declares the infinite to be an attribute of substance. The distinctive feature of this view is that one of the elements, for example, fire, or air, or water, is of some other nature than infinity, but that it has the attribute of infinity.

"The physicists, on the other hand, all of them, always regard the infinite as an attribute of a substance which is different from it and belongs to the class of the so-called elements—water or air, or what is intermediate between them."[81] In this view, the opposition between the primary nature and the other "things" of the world is understood as consisting in the eternity of the first nature, and the generation and decay of things; the element with the attribute of infinity is considered as "possessing" infinity, while the derivations, the phenomena, proceeding from the primary source, are limited and finite.

Aristotle adds to his list of the possible formulations, by reviewing that conception which declares the infinite to be that which is attributed to the elements in respect to their number. Anaxagoras and Democritus took the number of elements to be infinite, the former holding the position that the parts of this infinite number of elements are alike, while the latter held the parts to be different in size and shape.[82]

Against all these formulations, Aristotle opposes the following points:

(1) If the infinite is a substance which is neither a magnitude nor an aggregate, it will be indivisible (the Pythagoreans). But this sense of the infinite is irrelevant; just as the voice might be called invisible, in a sense, because by its nature it has nothing to do with being seen; the infinite refers to quantity, and to refer it to substance is going contrary to the facts.

(2) An actual infinite cannot have parts, that is, actual infinite parts—though in its real sense, as magnitude and quantity, it should have parts. (This is directed against the actual infinite as an attribute.)

(3) If there were an infinite number, it would be possible to traverse the infinite—which is impossible.

(4) The actual infinite is impossible; yet, in order not to deny his initial facts which demand some sense of the infinite—time, number, magnitude—the only other alternative must be accepted, and that is

that the infinite has a potential existence only. This potential infinite is not such that it will become actual, but a potential infinite which has as its characteristic note, succession; that is, "that is infinite, of which it is always possible, in regard to quantity, to take a part outside that which has already been taken."[83] In other words, there is an infinite, not where there is nothing left over and again over, but where there is always something left over.[84] The only possible infinite is that which is relative to this succession of moments within the magnitude or quantity which possess this potentiality.[85]

The position of Lucretius stands midway between that of Aristotle and that of Bruno. The difference between them serves to clarify the issue; for this difference becomes a pivotal point in the shift in conceptions as we shall see later. In this context the situation must be treated in its immediate connections. For Lucretius, it is certain that there cannot be any furthest point to anything; since there is nothing outside the sum of things, the sum can have no furthest point, and therefore lacks all bound and measure. It does not matter, then, in what part you stand or wherever you may be, the All is infinite in all directions.[86] In the infinity of space, there are infinite worlds; so that, when space is there, and if the selfsame force of nature remains, you must acknowledge that in other parts of space are other worlds and various tribes of men and races of wild beasts.[87] The universe is, hence, infinite; that this is so is deduced from the principle that that which is finite is limited, and that which is limited is looked at in relation to something else; consequently, that which is without limit, without boundary, must be infinite. Lucretius accepts the infinity of the universe, together with the infinity of worlds, existing in infinite space. In the first and fifth dialogues of "De l'infinito," Bruno employs similar passages to formulate his doctrine of the "extensive infinite," but transcends this teaching by declaring the infinity of the universe and space to be naught but the mirror of the infinite substance. For Aristotle, the infinite can only be predicated of that which is finite and determinate; it is only within a determinate subject that one can always take a part outside that which has already been taken. The division and the addition is not that which is infinite, but the continuous finite line, for example, is infinite with respect to addition or division. This applies to Lucretius directly, for the other and other space and limit of which he speaks merely mean, for Aristotle, that he is "extending" that which is still finite; but an extension of the sphere merely enlarges the "determinate," and it is within this determinate that infinity has meaning. If there is such an actual infinite in respect to addition, which

Aristotle has proved impossible, we would then have a subject infinite, and we would be exactly where we started.

From the fact that the infinite, for Aristotle, was a potential which could never be realized, and from the further fact that the infinite was confined to the determinate and finite, it followed that it could never be conceived as anything but an imperfection. But when the conception is transferred from the order of a qualitative aspect of being to the order of being itself, as was done by St. Thomas and Duns Scotus, the perfection of being "not only calls for all realizations, it also excludes all limits, generating thereby a positive infinity which refuses all determination."[88]

Infinity, in the sense of the "actual infinite," for St. Thomas, is that outside of which there is nothing; but because it is that outside of which there is nothing, the concept of infinity takes on for St. Thomas an entirely different aspect from that which characterized the concept for ancient philosophy. For St. Thomas, to prove the existence of an Infinite Being is to prove the existence of God; it is of the essence of God, and of God alone, to be infinite. To understand this concept, one must understand that it carries with it: (1) a stress on the transcendence of God with regard to the created world; (2) the radical contingency of all that is not God; (3) the actuality that is the perfection of God; (4) the exclusion of the concept of extension from the Infinite, because it is only the pure act of existing that is infinite.

> The perfection of the Christian God is that perfection, which is proper to being as being, that which being posits along with itself; we do not say that He is because He is perfect, but He is perfect because He is. And it is just that difference so nearly imperceptible at its point of origin, and yet so fundamental, that carries with it such startling consequences, when at last it brings forth from the very perfection of God, His total freedom from all limits and His infinity. . . . The divine being is necessarily eternal because existence is His very essence; He is none the less necessarily immutable, since nothing can be added or withdrawn from Him without destroying His essence simultaneously with His perfection . . . at the same time, because it is of being that God is the perfection, He is not merely its complete fulfillment and realization, He is also its absolute expansion, that is to say, its infinity.[89]

For Duns Scotus, the notion of the Infinite Being is not only the most fitting that we can form with regard to God, but the demonstration that the actual Infinite exists is also possible. Among other proofs Scotus employs that which rests on the infinity of intelligible objects: the divine mind knows all intelligible things; but since these are infinite in number, the divine mind, which is identical with the divine essence, must be infinite. Again, Scotus proceeds from identity of the divine essence and the divine perfections, to the infinity of the Supreme Being; since it is of the nature of an infinite substance to have its attributes coincide with its essence, whereas our human understanding is accidental to our being, it follows that the divine understanding is not a quality but is identical with the substance of the divinity.

The principle underlying Scotus' arguments is that God is perfect. For unless He were perfect, and unless He were infinite, He would not be the Being in which an intensive and positive infinity is identical with His nature. "For Duns Scotus, in fact, it is altogether the same thing to prove the existence of an infinite being, and to prove the existence of God, and this undoubtedly means, that until the existence of an infinite being has been established, it is not God whose existence has been proved."[90]

For medieval philosophy, then, the actual existence of an Infinite can mean only the existence of the Pure Being, God.

> We ought however, to add, that St. Bonaventure and St. Thomas are perfectly at one with Duns Scotus, in affirming the subsistence of a being, in face of whom absolute Heraclitianism and absolute Eleatism are equally vain, because at one and the same time, this being transcends the most intense actual dynamism and the most fully realized formal staticism. Even in the thought of those who are most attached to the aspect of realization and perfection which characterizes the Pure Being, we may easily discern the presence of the element of energy, which, as we know, is inseparable from the conception of act. For St. Thomas, as for Duns Scotus, it is of the very essence of God, as the pure form of being, to be infinite.[91]

Thus far, the notion of infinity has been shown to have been conceived in two principal ways. It was conceived by Aristotle as a potential infinity related to a succession of moments within a magnitude or a quantity: "A quantity is infinite if it is such that we can always take a part outside what has been already taken."[92] It was conceived in another way, by the scholastic theologians, as actual infinity pertaining to the essence of God alone; since it is of being that God is the perfection: "He is not merely its complete fulfillment and realization. He is also its absolute expansion, that is to say, its infinity."[93]

However, when we reach Nicholas of Cusa, there is a definite attempt to assign to the created universe a kind of infinity, while holding the conception that God alone is "positively" infinite. This is accomplished by asserting the "relative" or "privative" infinity of the universe.

For Cusanus, God is the "maximum absolutum," the greatest being, and at the same time the "minimum absolutum"; God is the "complication" of all things in the created world, and the world is the reflection of God. In relation to God, the world is finite; the universe is infinite only in the sense that it is the greatest of created things, and is thus a relative or a privative infinite. In spite of the fact that Cusanus declares the absolute to be both "maximum absolutum" and "minimum absolutum," and in a sense all that exists, he yet holds the universe to be the "maximum contractum," which is the finite image of the infinite God. The universe is thus a privative infinite as against the positive, perfective Infinite Being, God. God is perfect and absolute, and because He is perfect and absolute, He finds himself always in a state of perfect realization. The first principle is all that can be, and is of such nature that its actuality is identical with its power to be—that is, its power to be is always realized.[94]

Bruno asserts both the infinity of the universe and the infinity of Substance. Whereas Cusanus had called the universe the "explication" of the infinite God, Bruno calls the universe the "extensive" infinite as against the "intensive" infinity of Substance. The universe is infinite in extent, infinite in the number of its parts, and is constituted by elements which display a uniformity of Substance. The universe is the living reflection of the Infinite Substance; and while the infinity of Substance is all that can be, the universe is also all that can be, but with this major difference: each of the "parts" of the Infinite Substance is all that can be, whereas the parts of the universe are not all that can be.

> I call the universe infinite because it has no margin, limit, or surface; I do not call the universe totally infinite because each part that we encounter is finite; I call God all infinite because he excludes from himself every term, and every one of his attributes is one and infinite; I call God totally infinite because all of him is in all the world, and in each part of it totally and infinitely.[95]

Bruno's Infinite is without limit; it is that which is simply and absolutely perfect; it is that than which nothing greater can be; the measure of its possibility corresponds to the measure of its being; and in it, being, power, volition, and action are identical. Its nature makes it imperative that an infinity of things and world exist, in all possible modes; for if they did not exist, the primary conception involved in the necessity and perfection of an infinite and perfect Substance would be contradicted.

In our outline of the problem of the Infinite, we have traced the following steps:

(1) The attempt to make of the Infinite an actual Principle. Summarized, enlarged, and criticized, this led to

(2) The rejection of the Infinite, as a Principle, by Aristotle, and the affirmation of a potential infinite confined to quantity and magnitude.

(3) The concept of the infinite, as proposed by Lucretius, asserted the infinity of space, and the infinity of worlds within that space.

(4) In turning our attention to the medieval concept of infinity, as represented in St. Thomas, St. Bonaventure, and Duns Scotus, we found that they assigned the attribute of infinity to the being of God alone. We found further, in the philosophy of Nicholas of Cusa, the added attempt to assign to the created universe a kind of infinity, while still maintaining the positive infinity of God alone. And, finally, in Bruno, we find a new integration of all these conceptions.

The following chapters will show that the reconciliations achieved by Bruno coincide in a view which not only effects a higher synthesis, but also produces a formulation of the problem which is not, in principle, inconsistent nor contradictory.

Chapter I

UNIVERSAL SOUL AND UNIVERSAL FORM

"De la causa" is the product of Bruno's attempt to determine the constitutive principles of the universe. It is the result of an effort to analyze the structure of the universe in terms of its efficient, final, formal, and material causes. Among its chief topics are an examination of universal form and universal matter as distinct aspects of one eternal Substance; a discussion of matter in its dual role as subject and potency, culminating in the conclusion that actuality and potentiality are identical in the Infinite; an analysis of the principle of the coincidence of opposites and contraries, and of its application to the Infinite and its manifestations; and a demonstration that all diversity is nothing but a diversity of the aspects of one eternal and immutable Substance.[1]

That these goals can be achieved only through a limitation of the inquiry to those things which are within the range of our knowledge, Bruno indicates by his introductory remarks. If it is with the greatest difficulty that the First Cause and First Principle can be known, even in its traces, "how is it possible," he asks, "that those things which have a First and proximate Cause and Principle can be truly known, if their efficient cause, which is among those things which contribute to the true knowledge of things, is hidden?"[2] In other words, what are the foundations of our knowledge of things? And, can we allow ourselves to go beyond the limits of these foundations? The answer to both questions serves to define the scope of the discussion in the first dialogue.

Since the particular intellect has a determined relationship to forms abstracted from sensible things, it cannot know the First Cause through that form which is its own; this can only be known through the forms of its effects. Further, there are two kinds of effect: those which adequate the power of the cause, and through which the cause is fully known; and those which do not adequate the power of the cause and through which the nature of the cause cannot be fully known. Through this second kind of effect, it can only be shown that the cause exists; nevertheless, a cause is more fully understood from its effect the more perfectly the relationship is grasped. In this way, the cause is better known the more perfectly the productiveness is known. Likewise, the cause is better known the more perfectly the progression of the effect from its cause is known; for this manifests the similitude of the effect to its cause. It is to the second of these kinds of effect that Bruno turns in his formulation; for since our knowledge is initiated by dependent things, "We cannot infer other knowledge of the First Principle and Cause, than by the less efficacious method of traces." To know the universe, therefore, "is like knowing nothing of the being and Substance of the First Principle, because it is like knowing the accidents of accidents."[3]

Hence, although Bruno says that "it is better to abstain from speaking of so lofty a matter" and that it is "above the sphere of our intelligence," he yet urges a striving "toward the knowledge of that Principle and Cause," by allowing "the eyes of well-regulated thoughts to scan the magnificent stars and those luminous bodies . . . for they must have a Principle and a Cause."[4] With this statement, Bruno has reached his first goal, namely, a restriction of his subject· for it is obvious from the brief treatment which he has given of the incomprehensibility and the inconceivability of the nature of the First Principle and First Cause that he is reserving the resolution of this important problem until after he has discussed the First Principle and Cause "in so far as they are revealed in nature."[5] Here, he says, "We shall consider the Principle and Cause, in so far as in its traces it is either nature itself, or in so far as it reveals itself to us in the extent and lap of nature."[6]

Having thus set the stage, Bruno proposes two basic definitions on which he will build his conclusions. He defines *Principle* as "that which intrinsically contributes to the constitution of a thing, and remains in the effect"; and *Cause* as "that which contributes to the production of things from without, and which has its being outside of the composition."[7] By means of this distinction, Bruno introduces a conception which is of the greatest importance in the interpretation of "De la causa." "Principle" is not a concept distinct in essence from "Cause"; in fact, as he will point out later, both Principle and Cause are causes, the former receiving the name of Principle in order to distinguish its character of "immanence" from that of "Cause," whose nature it is to remain outside of the particular thing produced. "Principle" and "Cause" thus become two distinct terms, in order to stress the operational difference between an intrinsic and an extrinsic cause. That this is so is asserted by Bruno himself, when in attempting to support his characterization of the Universal Intellect as both Principle and Cause he describes it as "the true efficient cause, not only extrinsic, but also intrinsic, in natural things."[8]

A comparison of the two definitions bears this out. Since he has defined a "Cause" as "that which contributes to the production of things from without, and which has its being outside of the composition," the World Intellect is extrinsic Cause, because "as efficient, it does not form part of the things composed and the things produced."[9] And whereas he has defined a "Principle" as that which "intrinsically contributes to the constitution of things and remains in the effect," he will say, in speaking of the World Intellect as an intrinsic Cause, that "I call it an intrinsic Cause in so far as it does not operate around and outside of matter."[10] Thus the main distinction that Bruno wishes to make, by the terms "Principle" and "Cause," is between an intrinsic and an extrinsic Cause; he shows that the efficient Cause of the world is extrinsic, "by being distinct from the substance and essence of its effects, and because its existence is not like that of other things capable of generation and corruption." Although the world Cause, as intrinsic, works immanently in all, and in the particular composite, it works extrinsically with regard to the particular composition. This is to say, that in its dual capacity as Principle and Cause, the World Intellect exhibits two aspects of the same being.[11] This explains the previous statement concerning the diverse aspects under which God can be viewed—namely, that God can be called both Principle and Cause, as long as it is understood that He is One Substance, seen now as Principle, now as Cause.[12] Hence, as Principle, God is immanent in things, and as Cause He is distinct from things as "the producer from the thing produced." The distinction made above, therefore, with regard to the World Intellect, was made to point out the formal relationship that exists between God and the world of existing particular things. Further, when Bruno says that God is distinct from things as "the producer from the things produced," he does not mean to imply that God is completely transcendent, but that God, *as Cause*, is not exhausted by particular things; for were it to be understood that God is completely transcendent, His aspect as a Principle remaining in the thing would be neglected. This will be better understood by comparing these statements with those of Plotinus on the subject of the "presence" of an identical power in all things. If Bruno had adopted this conception of Plotinus, as some seem to believe, his conception of the inconceivability of God would also have conformed to that of Plotinus, according to whom the One is inconceivable because it is above nature. That this is not so, has already been pointed out above, where it was stressed that the entire opening discussion is directed by Bruno toward the investigation of God only in so far as He is revealed to us in nature. Throughout this discussion, the implication is felt that this is what he intends to show.[13]

We shall employ as our models for the Plotinian conception, Ennead iv.

3. 2-4 and Ennead vi. 4. 4. In Ennead iv. 3. 2-4, Plotinus explains the various ways in which it can be said that the "whole" and the "part" are related. Having stated that "there must be a soul that is not exclusively the soul of any particular thing, and those attached to particulars must so belong merely in some mode of accident,"[14] he asserts that "in such questions as this, it is important to clarify the significance of part."[15] Hence, he divides his inquiry into that which deals with "part" in relation to the "unembodied." He dismisses the first of these with the statement "it is enough to indicate that, when part is mentioned in respect of things whose members are alike, it refers to mass and not to ideal-form (specific idea)."[16] The second of these distinctions, that of part in relation to the "unembodied," he says, can be taken in various ways: "In the sense familiar in numbers, two, a part of the standard ten—in abstract numbers, of course—or as we think of a segment of a circle, or line (abstractly considered), or, again, of a section or a branch of knowledge."[17] The conclusion which is drawn in all these cases, whether it is with regard to the "material" or the "unembodied," is that part must be less than the whole, since they deal with quantity; but "since they are not the ideal-form Quantity, they are subject to increase and decrease."[18] Therefore, part cannot, in any of the senses taken above, be affirmed of the soul.

Well, then, "is it a question of part in the sense that, taking one living being, the soul in a finger might be called a part of the soul entire?"[19] The answer to this is the basis for Plotinus' resolution, for he says, "This would carry the alternative that either there is no soul outside of body or that—no soul being within body—the thing described as the soul of the universe is, nonetheless, outside the body of the universe."[20] Hence,

> If the particular soul is part of the All Soul only in the sense that this bestows itself upon all living things of the partial sphere, such a self-bestowal does not imply division; on the contrary, it is the identical soul that is present everywhere, the one complete thing, multipresent at the one moment: there is no longer question of a soul that is a part against a soul that is an all, especially where an identical power is present. Even difference of function, as in eyes and ears, cannot warrant the assertion of distinct parts concerned in each separate act—with other parts, again making allotment of faculty—all is met by the notion of one identical thing, but a thing in which a distinct power operates in each separate function. All the powers are present either in seeing or in hearing; the difference in impression received is due to the difference in the organs concerned[21] . . . The unit soul (it may be

conceived) holds aloof, not actually falling into body; the differentiated souls—the All Soul with the others—issue from unity, while still constituting, within certain limits, an association. They are one soul, by the fact that they do not belong unreservedly to any particular being; they meet, so to speak, fringe to fringe; they strike out here and there, but are held together at the source, much as light is divided on earth, shining in this house and that; and yet remain uninterruptedly one identical substance.[22]

Again, in Ennead vi. 4. 4, in speaking of the "Omnipresence of Authentic Existence," Plotinus says that diversity within the "Authentic" does not depend upon spatial separation, but upon differentiation. "The Intellectual Principle (which is Being and the Beings) remains an integral multiple by differentiation, not by spatial distinction."[23] Further,

> That principle distributed over material masses, we hold to be in its own nature incapable of distribution; the magnitude belongs to the masses; when this soul principle enters into them—or rather they into it—it is thought of as distributable only because, within the discrimination of the corporeal, the animating force is to be recognized at any and every point. For soul is not articulated, section of soul to section of body; there is integral omnipresence manifesting the unity of that principle, its veritable partlessness.[24]

Bruno's opposition is not to the content of these passages as such; nor is he opposed to the conclusions drawn from them by Plotinus. His dissatisfaction is not with what is said but with the fact that not enough is said. Everything that Plotinus has said applies to what Bruno has spoken of as "Cause"; and here there is agreement. What is left untouched is the additional treatment that Bruno has given to the second side of "Cause," which is its nature as an immanent Principle operating from within and "remaining within the composition." We shall see, indeed, that herein lies Bruno's radical departure from all those who assert the complete transcendency of a First Cause, and from those who hold that this First Cause is completely immaterial.

Thus far, Bruno's appeal has taken the following form:

(1) The goal has been to initiate an investigation into the constitutive principles of the universe, by declaring that God is both Principle and Cause. To make this meaningful, he has defined these terms, and has shown that they are distinct only in function. This has been accomplished by introducing, as an example, the Universal Intellect.

(2) The introduction of the "Universal Intellect" has served to lay

the groundwork for further applications of the concept of "distinction," which is to prove to be the clue to the major resolutions of the "De la causa"; for all distinctions which are made with regard to God, Principle, Cause, are distinctions between aspects of one and the same thing. Herein, lies the seed, which, when fully developed, will declare that all diversity and difference is not a difference and diversity of "Being," but of the manifestations and modes of "Being."

Returning to our text, we find that Bruno at this point further defines his intention; for, since he has treated Principle and Cause in their formal aspects, his next step is to apply these concepts to the world of particulars. That he sees fit to do this, is further evidence of the fact that he realizes that the more he knows of the actual operations of Principle and Cause, the more certain will be his knowledge of their formal constitution. The goal is reached in the following ways:

(1) Bruno employs the Aristotelian four causes, but he superimposes his own conceptions of Principle and Cause upon them, and restricts their application to the universe as a whole, rather than to the particular as particular. This permits the introduction of the "new" term "World Soul," meaning the Universal Form which is viewed through its principal faculty, the Universal Intellect.

(2) In the end, this World Soul or Universal Form turns out to be that which is being investigated by means of these causes. Thus the conclusion will stress the fact that "Form" is one of the attributes of God. We have arrived at that attribute through a study of:

(*a*) the higher distinction of Principle and Cause
(*b*) the lower distinction of Universal Intellect and Universal Form
(*c*) the still lower distinction of the Universal Intellect and its operations; that is, the observational distinction gathered through watching it as it operates either as a formal cause, or a final cause, or as an efficient cause.

(3) All of the above is drawn together and expressed by this general thesis: all diversity and difference is a diversity and difference due to these ways of approaching that which is not diverse and different—namely, God and His attributes.

Bruno indicates this order and connection of causes, where he says: "and as to causes, I should like to know first about the First efficient Cause, about the formal Cause which you say is joined to it, and then about the final Cause, which is understood to be the mover of the latter."[25] The Universal Intellect is the universal efficient Cause; it is "the most intimate, the most real, and the most proper faculty and partial power of the World Soul."[26] It is both a Principle and a Cause: "It is one and the same thing which fills the whole, illumines the universe, and directs nature to produce its various species, as is fitting."[27] By thus describing the functions of the Universal Intellect, Bruno clearly shows that, whereas he has called it the efficient Cause, this does not mean that he is neglecting its correlative function.[28]

In order to show that the formal Cause is joined to the efficient Cause, Bruno draws the analogy between artificial things and the products of the Universal Intellect; for, if the former are not produced without reason, "how much greater, then, must we believe to be that creative intellect which is not restricted to one part of matter, but as a whole operates continually throughout the whole."[29] And because no agent that works according to the rules of the intelligence will be able to obtain effects, unless they are in accordance with some intention, so likewise is it with the Universal Intellect, because "it is necessary that it possess them all in advance, according to some formal concept."[30] And because the Intellect takes such pleasure and delight in calling forth all sorts of forms from matter, the aim and final Cause of the Universal Intellect is the perfection of the universe, "which means that all the forms are actually existent in the diverse parts of matter."[31]

Though the terms employed are those of Aristotle, it is obvious that Bruno has radically changed their meanings: whereas Aristotle confined the analysis of the four causes to particulars, Bruno applies the scheme of the four causes to the universe as a whole. In Aristotle, the final cause of natural processes is immanent in the thing produced, in the sense that it is the form that is coming to be: the "what" and the "that for the sake of which" are one.[32] This immanence, for Aristotle, is what distinguishes final causes in nature from those in art, "for it is plain that this kind of cause is operative in things which come to be and are by nature." Hence, "since nature means two things, the matter and the form, of which the latter is the end, and since all the rest is for the sake of the end, the form must be the cause in the sense of that for the sake of which."[33] In other words, we can say that for Aristotle, besides the efficient, final, and material causes, there must be an ultimate stage in the coming to be; this is the actuality and the form. Again, the efficient cause in Aristotle is external and antecedent to the thing produced, "either as the primary source of the change or coming to rest,"[34] or, "as the answer to the question, what initiated the motion."[35] But for Bruno the efficient Cause works from within, remaining distinct only because it is not exhausted by the particular thing it produces; and the final Cause is that which assures "the perfection of the universe."

In Plotinus, too, the similarity in terminology is a sufficient basis for comparison. While Bruno holds that the formal reasons are the "preconceived" ideas in the Universal Intellect, and hence the model of the action of the efficient Cause, Plotinus holds that the soul and its activity are directed by the contemplation of the ideas in the Universal Intellect.[36] Two formal distinctions must be made at this point: (1) For Plotinus, the Universal Intellect and the World Soul are separate, while for Bruno, the Universal Intellect and the World Soul (Universal Form) are one. (2) For Plotinus, there is a twofold distinction, based upon the relationship between the Ideas as they exist in the Intellect and their existence in the material forms which are derived from them; in the former case, the Ideas are not distinct from each other, in a real sense; in the latter case, there is a real distinction. This serves to emphasize the fact that for Bruno, the ideas constitute the "plan" of the universe; though existing formally in the Intellect, they are not really distinct in their existence, but they penetrate all things in such a way that they are not really distinct, either with regard to their relationship among themselves, or with regard to their relationship amongst all existing things.[37] Therefore, though we may say that the forms correspond to the ideas, it is better to say that the correspondence is based on an identity.[38]

The Plotinian separation of Intellect and Soul serves not only to introduce Bruno's next point, but also to clarify it. This is made evident where Bruno indicates that from the treatment of the principal faculty of the World Soul, he has gained sufficient material with which to approach the treatment of the World Soul itself.[39]

To the vertical generation of hypostases in Plotinus, must be added the conception of diversification within the hypostases. Thus, to understand the statement concerning the separation of Intellect and Soul, it is not enough to point out that the higher hypostasis generates the lower, nor to say that the higher remains separate from the lower, nor to say that the hierarchy of hypostases in Plotinus is made up of the One, Mind, Soul, Sense, Nature, and Body. To this must be added the fact that, in Plotinus, vertical generation of hypostases is supplemented by a horizontal diversification within each one of the hypostases. Hence, "Soul, for all the worth we have shown to belong to it, is yet a secondary, an image of the Intellectual Principle. . . . What the Intellectual Principle must be is carried in the single word that Soul, itself so great, is still inferior."[40] And:

> The unit soul (it may be conceived) holds aloof, not actually falling into the body; the differentiated souls, the All Soul, with the others, issue from the unity, while still constituting, within certain limits, an

> association. They are one soul by the fact that they do not belong unreservedly to any particular being; they meet, so to speak, fringe to fringe; they strike out here and there, but are held together at the source, much as light is a divided thing upon earth . . . and yet remains an identical substance.[41]

The first of these differences applies to our present problem; the second will prove invaluable in our discussion concerning the relationship between "particular souls" and the World Soul. As we shall see, Bruno's answer to both problems is essentially identical.[42]

It now becomes apparent that Bruno, in treating of the Universal Intellect, was, in fact, treating of the World Soul. Hence, the World Soul, in so far as it animates and informs things, is an intrinsic and formal part and constituent of the universe; as such it operates as a Principle.[43] But in so far as it directs and governs (by means of the Universal Intellect, in its capacity as efficient, final, and formal Cause), it is not a Principle, but a Cause. Even Aristotle, says Bruno, has so considered the intellectual soul, "understanding it as an efficient cause, separate according to existence from matter, saying that it is a thing that comes from outside."[44] That Bruno quotes this passage at this point, is significant only because he means to apply it to the "World Soul"; this is indicated where he says: "And if this intellectual power has the functional role of an efficient Cause, much more should we say the same thing of the World Soul."[45] Bruno's agreement with the Aristotelian text which he is here citing finds its reason in the fact that he is seeking support for his own position; hence his agreement is not with Aristotle's definition of "cause," as has been shown above, but with Aristotle's description of the intellect as an efficient cause, understood in the sense of a principle of operation which is not solely dependent upon matter for its entire activity. Bruno, therefore, is not only interpreting the text from his own point of view, but also superimposing his own meanings of Cause and Principle on that text. This becomes evident from an analysis of Aristotle's statement. When Aristotle says that the intellect is a thing that comes "from without," he is seeking to affirm that while a form which has no operation apart from its union with matter exists by the same act as that with which it is combined or mingled, since it possesses the same existence, a form like the intellect, which is not totally educed from matter, or whose existence is not solely directed towards that concretion, exists by reason of an extrinsic principle. It is in this sense that, for Aristotle, the intellect comes from "without." When Bruno says "from without," he means that the efficient Cause is "without" the particular composite and not exhausted by it.

In Bruno's text, the attempt has been to show these things:

(1) That the World Soul, as Principle, is the formal constituent in all things, "because spirit is found in all things, and there is not the least corpuscle that does not contain in itself some such portion that spirit would not animate."[46]
(2) That the World Soul, viewed through its principal faculty, the Universal Intellect, is the efficient Cause of all that is
(3) That the Universal Intellect, as efficient Cause, is not to be confused with the things and composite beings which it produces
(4) That the World Soul and the Universal Intellect work from within; hence they are intrinsic Principles
(5) That "Spirit," "Universal Form," and "World Soul" are synonymous terms: This is indicated where Bruno says that he agrees with Anaxagoras, "who held that everything is in everything because, since spirit, or soul, or universal form, exists in all things, all can be produced from all."[47]
(6) That the progression has been from God as Principle and Cause to Universal Intellect as Principle and Cause, to World Soul as Principle and Cause, to Universal Form as Principle and Cause.[48]

In stating that the World Soul, as Spiritual Substance, "is the formal constitutive Principle of the universe,"[49] Bruno is not making spirit, or soul, or form, superior to matter, but is restricting himself in the present context to the aspect of form in its nature as Cause and Principle. But he implies that matter is also to be posited as a substantial Principle, when he says that although form is one in all things, it is "however, according to the diversity of dispositions of matter, and according to the power of the material principles, active and passive, that it comes to produce diverse configurations and to effect different powers."[50] Hence, he implies that the Formal Principle, in its diversification, is in some way dependent upon matter; and, in so doing, he lays the groundwork for his future assertion of the latter's substantiality. At this point, he reverts to the questions which he had left undecided—namely, the significance and relationship of particular souls to the World Soul.

All things, then, that are particular, are compositions of both the formal and material Principles contracted to be this particular thing; and so, "while this form changes place and condition, it is impossible that it be annulled, because the spiritual Substance is not less subsistent than the material."[51] Only external forms change because they are not things, are not substances, but are "of Substance and about Substance."[52] In other words, particulars are "accidents and circumstances of substances"; whereas nature assures us that "neither bodies nor the souls should fear death" because matter and form are constant Principles. In this last statement, Bruno attempts to counter the Aristotelian argument which holds that the form is the act of the body; for if the body is corruptible, and the form is not, and they are joined essentially, the dissolution of the body would (according to Bruno's interpretation) involve the dissolution of its essential concomitant, the form. This, of course, leaves out the consideration that Aristotle had given concerning the intellectual soul in holding that there is a principle of operation in the intellectual soul which does not depend completely on matter. But Bruno, who is directing the discussion toward the consideration of his principles, is consistent in that he has turned the discussion to the position which will prove his own definitions to be valid. He had asserted that both form and matter were substantial Principles; hence, a dissolution of Principles would entail a dissolution of being, and this is the impossibility which is thus apprehended.

"Particular" souls are, therefore, "modes" of the World Soul.[53] Thus, whereas Plotinus would describe the Universal Soul as a hypostasis which, without diminishing itself, engenders the particular souls,[54] Bruno declares the Universal Soul to be the intrinsic and formal constituent of particular souls, making the latter no more than particular modes or accidents of the Universal Soul.[55]

It is the same World Soul which is present throughout all particular things. Individual souls are not subsistent realities, but multiplied aspects of the one Universal Soul. Whereas, for Plotinus, the individuality of the particular souls is respected,[56] this is not the case for Bruno; the unity of the Universal Soul is not for him a generic unity.[57] For Plotinus, particular souls are as subsistent in their unity as is the World Soul; this is based on the Plotinian principle of the separation between the One and the Intellect, between the Intellect and the Universal Soul, etc., so that one engenders the other without defect to itself, or without being absorbed in the effect.[58] In this way, when Plotinus says that all souls are one soul, he means to say that this is so because they proceed from one soul.[59]

Bruno's position concerning the relationship between the Universal Soul and "particular" souls is further clarified by his desire, now made evident in the text, to show that the different kinds of forms do not deserve the rank of Substance; hence, though various kinds of form do exist, they are all "contractions" and "manifestations" of one form. It is not the Universal Form that is different in the particulars; this "difference" that is seen is dependent upon the following circumstances: Form can exist (1) as it communicates the perfection of the whole to the parts, remaining extended, dependent, and without its own activity (material form); (2) according as it informs and is dependent but unextended, perfecting and activating the whole, while remaining in the whole and in every part (vegetative and sensitive soul or form); and (3) according to the way it activates and perfects the whole, but does not remain extended, and is not dependent with regard to its operations. In this way, it exercises intellectual powers, and does not form any part of man that can be called man (intellectual soul).[60] All these ways of manifestation do not contradict Bruno's statement, that the Universal Form is "invariable in itself"; this is made clearly evident where he says: "Though invariable in itself, it is yet variable through particular subjects and the diversity of matter."[61] This is one of the ways in which form and matter determine each other.

> Form and matter determine each other because the form, having in itself the facility to constitute particulars of innumerable species, comes to contract itself to constitute one individual; and, on the other hand, the potentiality of indeterminate matter, which can receive any form whatsoever, comes to be determined into a species; the one is the cause and definition of the other.[62]

With this statement Bruno has reached the end of the first dialogue. This brings him to the second aspect of the problem of form and matter; that is, the problem of form and matter viewed this time from the aspect of matter, as the potentiality of all that exists.[63]

Chapter II

THE MATERIAL PRINCIPLE AS POTENCY

HAVING INDICATED that both form and matter are Principles because the one is the cause of the definition and the determination of the other, Bruno proceeds to an analysis of matter, in much the same manner as he had previously treated form. He confesses that he at one time held that matter alone was a substantial Principle: "And for a long time, I myself had been an adherent to this conception." But that he found himself unable to remain in the company of those with whom he had allied himself, he indicates where he says: "We find it necessary to recognize in nature 'two kinds' of Substance, one which is form, the other which is matter."[1] This is based on the fact that "it is necessary that there be a most substantial act, in which there is the active potency of everything; and also a potency and a subject in which there is equally the passive potency of all; for in the former is the power to make, and in the latter is the power to be made."[2] For Bruno, it seems obvious that no philosopher can posit an active Principle, such as the Universal Form or World Soul, without at the same time positing its correlative, a passive principle. It is inconceivable to Bruno that there should be a World Soul which orders and brings into being the things of the universe, without the positive existence of that from which, and in which, it can exercise and produce its effects.[3] Thus the two, matter and form, accompany each other; they are necessary correlatives, and they are two aspects of a single Substance which is both.

Having established the principle, Bruno realizes that the "de facto" proof is wanting. This he initiates by employing the analogy of art, which he admits has had a long history with regard to "demonstrating the existence of matter."

> All these arts bring into a particular matter diverse pictures, arrangements, and figures, none of which is proper and natural to that matter; and, therefore, nature, to which art is similar, needs to have for its operations a matter, because it is not possible that there be an agent which, when it wishes to make something, does not have that out of which it can be made.[4]

This last statement is a particular application of the formal principle enunciated above—that is, "that there is then a kind of subject from which, with which, and in which, nature effects its operations and its work."[5] Just as in the analogy of art, wood as wood has no artificial form, but can have any and all forms through an agent working on the surfaces, just so prime matter, as a Principle, has no natural form, but can have any and all forms through its agent, which is its correlative active Principle, the Universal Form.

From the formal analogy, which stressed similarity, Bruno draws the basic differences between artificial matter and natural matter. Whereas he had pointed out that both possess, albeit analogously, an active agent and a subject, still, in the case of artificial matter, the agent works only on the surfaces of matter, "a thing already formed by nature"; while, on the other hand, in natural matter, the agent works as it will, from within its subject or matter, "which throughout is formless."[6] Again, whereas the subjects of art are many, the subject of nature is one; for while the former are diversely formed by nature, and hence "different and variegated," the latter, being formless, is indifferently all forms, "since all difference and diversity stem from form." Lastly, artificial and natural matter are known in a different way; for whereas artificial matter is "seen with the sensible eyes," natural matter can be "seen only with the eyes of reason."[7]

For Plato, who does not employ the term, "matter" is spoken of as the eternal "receptacle" of all things; and since it receives all species, it must, for Plato, be deprived of any form of its own. It is the eternal "place," the eternal theatre of generation and corruption. It possesses no form, nor is its being accessible either to sense or to reason. It receives all forms within itself, without becoming any particular form, for it is the common "place" and the common "receptacle" of all forms. This is evident where Plato says:

> In the same manner, we should speak concerning that nature which is the general receptacle of all bodies; for it never departs from its own proper power, but perpetually receives all things; and never contracts any form in any respect similar to anyone of the intromitted forms. It lies indeed in subjection to the forming power of nature, becoming agitated and figured through the supernally intromitted forms; and through these it exhibits a different appearance at different times.[8]

The receptacle, therefore, is neither accessible to sense nor to reason; its "existence" is affirmed as a result of what Plato calls "illegitimate reasoning"[9]; it has neither any other movement nor any other form than the movements and forms which it contains.

For Aristotle, "prime matter" is not known except by means of analogy; "for as the bronze is to the statue, the wood to the bed, or the matter and the formless before receiving form to anything which has form, so is the underlying nature to substance, that is, the 'this' or existent."[10] Matter is what subsists through the generation and corruption of things.[11] For Aristotle, matter is not a reality in itself; for the material subject never exists separately without an act, and is known only through its relation to the act which it receives from a particular form, which is its correlative.[12] "Prime matter" is something which can enter into the constitution of being, but it itself is not being. Matter is in itself wholly indeterminate because neither by its notion nor in virtue of its own intelligibility does it require this form rather than that form; this is so because in itself matter has no notion, being, or intelligibility.

For Plotinus, matter is not, as in Aristotle, correlative to a particular form; it is absolute privation, absolute non-being. Although he employs a method which is analogous to Plato's, Plotinus arrives at a "what" which is not itself the eternal spatial receptacle of Plato, but an absolute bare substratum—in which the multiple forms are born—which receives extension, but is yet not extension.[13] Like Plato, Plotinus arrives at his conception of matter by means of a "spurious reasoning"; this is manifest where he says: "The representation of matter must be spurious, unreal, something sprung of the Alien, of the unreal, and bound up with an alien reason."[14] For Plotinus, matter is neither body, nor soul, nor intelligence, nor life, nor form, nor reason, nor limit.[15]

For Bruno, matter is not non-being; it is not unintelligible in itself. In its own right it is an intelligible Principle, on an equal level with the formal Principle, the Universal Form. It is not the subject of particular determination, as it is for Aristotle; it is not a "prope nihil," nor a privation, nor non-being; it is as real as existence itself. Matter is viewed by Bruno as the second of the two constitutive principles of all things.[16] This does not mean that there are two separately existing principles; there is no real distinction between these two principles; they are two aspects of one Substance which is both. "And verily, it is a necessary thing that just as we can posit a constant and eternal material Principle, so also should we posit, in a similar way, a formal Principle."[17] This statement is vital for four reasons: (1) It indicates that both the "form" and the "matter" are substantial Principles; this, in itself, has been shown to be the case in what has preceeded. (2) The important declaration is the fact that Bruno has restated his position of the first dialogue, this time, however, placing the material Principle first. (3) He restates his conclusion concerning the relationship between particulars and the

Universal Form, this time from the point of view of matter; for particulars "have no being without matter, in which they are generated and corrupted, and out of whose bosom they spring and into whose bosom they are taken back."[18] Particulars, in short, "are not Substance, or nature, but things of Substance and things of nature."[19] (4) He applies the new findings to the particular things from the point of view of their constituents: "Therefore, we say, that there are three things in this body: first, the Universal Intellect inherent in all things; second, the unifying soul in all; third, the subject."[20] In other words, the first two constitute the Universal Form, and the third is the material Principle, as subject.

To return to our text: Bruno, at this point, indicates that though he has been concentrating his attention on the aspect of potentiality, there is yet another side of matter which he shall have to treat in detail—namely, matter as subject: "This Principle, which is called matter, can be considered in two says: first, as potency; second, as subject."[21] This statement serves to introduce another distinction, that between active and passive potency. But this is only another way of treating the material Principle, depending upon the aspect from which it is being investigated: "If there always has been the capacity to make, to produce, to create, there always has been the potency to be made, to be produced, to be created, because the one necessarily implies the other."[22]

These conclusions are closely allied to those of Cusanus. Cusanus, too, would hold that there is no pure act opposed to pure possibility. Matter thus becomes, for Cusanus, something that does not exist outside of God, "for that which exists was possible"; and matter is, therefore, the "possibility of being."[23] Absolutely considered, according to Bruno, act and potency are one "because the absolute possibility, through which the things are in act can be, does not exist before that actually, nor after it; moreover, the capacity to be exists together with the being in act, and does not precede it; because if that which can be could make itself, it would be before it was made."[24] When we apply this maxim to the Absolute Principle, says Bruno, the truth becomes evident; for if this First Principle is all that can be, "it itself would not be all, if it could not be all; in it, therefore, the act and the potency are the same thing."[25] The highest Principle is all that can be, and contains in itself all being, "because it is all that any other thing is, and can be any other thing that is or can be."[26] In particulars, the potency is not equal to the act, because the act is limited and not absolute: "Every potency and act, which in the Principle is enfolded, unified, and one, is in other things unfolded, dispersed, and manifold," because particulars never have more than a single specific and particular being, and when the particular "refers itself to every form and act, it does so through certain dispositions, and through a certain order of succession of one being after another.[27] The universe, which is the image of the First Principle, is also all that can be, yet in a different sense; for although the universe is all that can be, no one of its parts is all that can be. The universe is all that can be, "in a developed, dispersed, and distinct way," while the First Principle is all that can be, "unifiedly and indifferently; it is all in all, and the same in all, simply and without distinction and difference."[28]

Through showing that matter can be viewed either as a potency or as a subject, and following this with the further distinction between active and passive potency, Bruno has not only indicated his conception of these determining Principles, but has also had an opportunity to introduce his distinction between God as First Principle and the universe. As we shall have occasion to see later, the conception of infinity is based upon this distinction. Here, Bruno is not prepared to deal with this vital problem; he is content to present his thesis piecemeal, and topic by topic, until that moment when he can say that everything which has been presented to you, in these dialogues of "De la causa," forms an integral part of the infinity of the First Principle, God, or Substance.[29]

For Bruno therefore, as for Cusanus, God is absolute potency and absolute act. From the consideration of this identity in God, Bruno has drawn the following conclusions: first, that matter is not to be understood as something outside of God, the First Principle, but is to be conceived as one with the nature of the First Principle; second, that this material Principle is the correlative of the formal Principle—"whence it is to be understood, that there is in the universe, a First Principle, no more distinctly material than formal, which can be inferred from the aforesaid similitude between absolute potency and absolute act"[30]; third, that whereas Cusanus holds the universe to be a relative 'infinite,' a finite image of the infinite God, Bruno, basing himself on the coincidence of potency and act, concludes that the universe is the infinite image of the Infinite First Principle[31]; fourth, that these distinctions between potency and act, between matter as potency and matter as subject, are, alike with those of Universal Soul, Universal Form, and Universal Intellect, distinctions of one and the same eternal Substance.[32]

Chapter III

THE MATERIAL PRINCIPLE AS SUBJECT

In the third dialogue Bruno distinguished between matter as potency and matter as subject. Having considered the former of these distinctions, he proceeds in the fourth dialogue to treatment of the latter. This is initiated by introducing the concept of the intelligibility of matter. By demonstrating that matter is intelligible, Bruno adds to his mounting list of reasons in support of his contention that matter is a substantial Principle. The goal is to transfer the notion of matter beyond the limits of the spatial world; once this is achieved, he will have demonstrated that everything is composed of both matter and form.

Because the analogy of art had proved invaluable in gaining the understanding of the notion of "prime matter," Bruno again applies it here. "Surely, it cannot be denied that just as everything sensible presupposes the subject of the sensible, so also everything intelligible presupposes the subject of intelligibility."[1] Since there are two subjects, a further application of the analogy should show that there is something common to both: "It is necessary, then, that there be something that corresponds to the common concept of one and the other subject."[2] That "common" something must, therefore, be the material Principle.

The notion of intelligible matter is a difficult one. Realizing this, Bruno admits that he relies on Avicebron and Plotinus for support. The inverse order is not accidental; Avicebron takes precedence in this instance, for it was his teaching concerning the notions of Universal Form and Universal Matter which gave impetus to Bruno's conception of Form and Matter. Bruno's conception, therefore, will be better defined if it follows a brief summary of Avicebron's principles.

Avicebron held that all separate substances are composed of both matter and form[3]; for if they were not so constituted, there could be no diversity among them. "To be a subject" belongs to the concept of matter, but matter is made diverse through form. If, therefore, these separate substances were matter alone, diversity would be missing; and if they were form alone, diversity would occur only if they were a "subject"; but this has already been shown to be a characteristic of matter and not a characteristic of form. No created substance, therefore, can be a pure unity; so that all substances, except the uncreated one, are a combination of matter and form; matter, then, is universal. We can trace the following properties which are constant throughout its range: (1) it has a single essence; (2) it supports diversity; (3) it includes, as a subspecies, corporeal matter; (4) it is one of the two principles which are the roots of all created beings.[4]

In Ennead ii. 4. 1-5, Plotinus reveals his conception of intelligible matter. He initiates his treatment by marking off the intelligible from the corporeal; the latter is the common matter of particular corporeal things. The steps by which Plotinus achieves the development of his conception are: (1) it must be shown to have a positive existence; (2) this will lead to the distinction between itself and the common matter of corporeal things; (3) by marking off its characteristics, its real place can become known; and (4) this is its place on the rungs which mark off the hierarchy of existence.[5]

In Ennead ii. 4, 1, after speaking of that school which makes matter the stuff underlying the primal constituents of the universe, Plotinus declares that there is still "another school which makes it incorporeal . . . and some of them, while they maintain the one Matter, the foundation of bodily forms, admit another, a prior, existing in the divine sphere, the base of the Ideas there, and of the unembodied things."[6] Just as there is a common matter for things corporeal, there must be a common matter of things incorporeal; for "we are obliged both to establish the existence of this other kind, and to examine its nature and its mode of being."[7] Hence, "the matter of the realm of process ceaselessly changes its form; in the eternal, Matter is immutably one and the same, so that the two are diametrical opposites."[8] With regard to the former, he says that "the matter of this realm is all things in turn . . . so that nothing is permanent and one thing pushes another out of being; Matter has no identity here.[9] Of the matter of the intelligible realm, he states "that in the Intellectual it is all things at once, and, therefore, has nothing to change to; it already and forever contains all."[10] Intelligible matter, then, is the only "kind" that can have "real being" attributed to it because it alone is that to which can be attributed the "Base in Being." It can then be distinguished in accordance with the rung it occupies on the ladder of being;

> There is, therefore, a Matter accepting shape, a permanent substratum . . . the matter of this sphere is not living or intellective, but a dead thing decorated; any shape it takes is an image; exactly as the Base is an image . . . There, on the contrary (the intelligible) the shape is a real existent, like the Base . . . the Base there is Being.[11]

For Bruno, whose conception of intelligible matter reflects the views of Plotinus more than those of Avicebron, in spite of Bruno's own emphasis on the latter, the concept of matter belongs to incorporeal as well as do corporeal things because "the former have their existence through the capacity to be, no less than the latter have their being through the power to exist."[12]

From the application of this principle arise three further conclusions: (1) incorporeality is not to be confused with immateriality; (2) incorporeality is the correlative of corporeality; (3) immateriality is, accordingly, a contradiction in terms.

> Corporeal matter exists in virtue of the dimensions, and the extension of the subject, and those qualities which have their existence in quantity; and this is called corporeal substance, and presupposes corporeal matter . . . or, it exists without these dimensions, extension, and qualities; and this is called incorporeal substance, and similarly presupposes the said matter.[13]

These distinctions are now reapplied to the formal distinctions made above between active and passive potency, and serve as natural properties from which Bruno draws further conclusions concerning the essence of the material Principle: "To an active potency, common as much to corporeal things as to incorporeal things . . . there corresponds a passive potency, as much corporeal as incorporeal, and a capacity to be, as much corporeal as incorporeal."[14] Thus, while in the eternal there is a matter which is always under a single act, there is in the variable, and mutable, a succession of acts. In the eternal Principle, matter has all that it can have, since its absolute potency is identical with its absolute act; it is all that can be, at once, always and together. But in the variable and the mutable, where potency is limited by act, and act by potency, matter has not all that it can have; what it does have, it has not at once, but at different times, and in a certain order of succession. Further, the matter of the First Principle, "through being actually all that can be, has all measurements, has all the species of figures and dimensions; and because it has all of them, it has none of them; because that which is so many different things need not be any one of those particular things."[15] Matter as Principle, then, is not really distinct from Form: "Matter is at the highest grade of purity, simplicity, indivisibility, and unity because it is absolutely all; for if it had certain dimensions, a certain being, a certain figure, a certain property, and a certain difference, it would not be absolute, and would not be all."[16]

From the identification of Form and Matter, and from the demonstration that potency and act, absolutely considered, are one, Bruno draws the following conclusions: (1) There is in the universe a scale of being similar to the "capacity to be"; "just as the formal concept ascends, so does the material concept ascend." (2) If measurability is taken as an aspect of matter, it is obvious that the material Principle is absolutely free from dimensions; as absolute, matter is above all and comprises all, though, as contracted, it is comprised by some and under some. Matter in itself has no determinate dimensions, and for that reason is understood as indivisible, and as receiving dimensions by virtue of the form it receives: "It has some dimensions under the human form, others under the form of a horse . . . and for this reason, before it exists under any of these forms, it has in potentiality all those dimensions, just as it has the potency to receive all those forms."[17] (3) "Matter contains the forms and implies them"; matter, which unfolds all that it possesses as enfolded, must be called the "divine and excellent progenitor and mother of all things."[18] (4) Matter, in itself, receives no perfection from the particular form which it possesses at any particular moment.[19] (5) Matter does not receive any form from without, but "sends forth the forms from her bosom, and consequently has them in herself." Matter is eternally all forms, because it is one with the Universal Form.

How then do these forms come to matter? They are certainly not to be understood as Plato conceived them, as copies of eternal Ideas in which they participated in some inexplicable way.[20] Particular forms are accidents of matter, which have their true being within that matter "in itself." Matter is the constant and eternal source of all forms. The diversity of genera, species, properties, alterations, and changes is not a diversity of substances, but of appearances of one unique, immobile Substance.[21] The totality of beings is reduced to one Being, which is divine, infinite, and immortal.[22] Particular things have their base in the real, in absolute Form and absolute Matter; they are modes of one eternal Substance, which is both form and matter. Substance is dynamic, producing a multiplicity of things; the unfolding of the enfolded proceeds from the unity of Substance to multiplicity and back. All things are composed of form and matter, not the particular form of Aristotle, but the absolute form which is one with absolute matter.[23]

Chapter IV

SUBSTANCE AND INFINITY

In the fifth dialogue of "De la causa," Bruno effects a summary of his conclusions. This summary not only serves to clarify the issues, but also to develop them into a well-organized unit. Taken together, these conclusions form the body of Bruno's conception of Substance and Infinity. The attempt here, as in the three preceding dialogues, is to describe the notions which have been drawn from the analyses without recourse to the problem of definition.

In the course of this summary, it will become increasingly evident that the most important conclusions are a direct result of Bruno's application of the concept of distinction. Throughout the investigation, there has been the constant reminder that all results are dependent on this principle. It is so important, that it is safe to make this declaration: that unless one knows how to distinguish without separating, Bruno is forever lost to the understanding.

The points which are taken up in the order of their appearance are:

(1) The universe is declared to be one, infinite, and immobile.

(2) Substance, because it is one, cannot be generated or corrupted.

(3) Substance, which is both Matter and Form, is neither.

(4) In Substance, there are no parts; there is no difference between part and whole in the Infinite.

(5) In the Infinite, the indivisible does not differ from the divisible.

(6) Substance, or the Infinite, is all that can be; "particulars" have all being, but they do not have all the modes of being.

(7) There is in the universe a scale of being, through which nature descends to the production of things, and the intellect ascends to the cognition of them.

(8) All contraries coincide in the One.

From the fact that every distinction is one of aspect, Bruno declares the universe to be "one, infinite, and immobile. One, I say, is the absolute possibility, one the act, one the form or soul, one the matter or body, one the thing, one the being, one the greatest and the best." The Infinite "is without end or limit . . . and insofar infinite and indeterminate . . . and consequently immobile."[1]

This unique substance, because it is one, infinite, and immobile, is not generated nor corrupted; for "there is no other being which it might desire or look for, since it has all being"; and "there is no other being into which it could change, since it is itself everything."[2] There is no "other" for the Infinite, for what can be "other" to that which is all?

Because Form and Matter are distinctions of one and the same Substance, this Substance is both and neither: "It is form in such a way that it is not form; it is matter in such a way that it is not matter; it is soul in such a way that it is not soul; because it is all indifferently, and therefore one."[3]

Because the Infinite is one, it can have no "parts"; there is in it no difference between part and whole; there is no greater and lesser: "because a greater part does not conform more to the proportion of the Infinite than any other smaller part; therefore in its infinite duration, the hour does not differ from the day, the day from the year, the year from the century, the century from the moment."[4] Everything is as much substance as any other thing. "Thou canst not more nearly approach to a proportion, likeness, union, and identity with the Infinite by being a man than by being an ant; not more nearly by being a star than by being a man; for in the Infinite these things are indifferent."[5]

Since the Infinite is indifferently all, "the indivisible is not different from the divisible, the simplest from the infinite, the center from the circumference."[6] Where all is one, and all is substance, we cannot say that the line, the point, the center, the circumference, the maximum, the minimum, the finite, the infinite, are different.

If all things are indifferently one thing, what reality can we give to the particular things in the universe? And this is the difference between the universe and the things of the universe; the universe comprises all being and all the modes of being; of particular things, each one has all being, but not all the modes of being."[7] Whereas the Infinite includes all being totally, each particular includes all being, but not totally; in the former, there is nothing outside it; in the latter, there is always something outside and beyond: "Therefore, it is to be understood that all is in all, but not totally, and in all modes in each one; understand, therefore, that everything is one, but not in the same mode."[8] Multiplicity results from a multiplicity of modes of being, which is one; multiplicity and difference is pure accident, pure figure, and pure alteration of Substance; all things are in the universe and the universe is in all things, "because all things concur in a perfect unity"; the universe is one everywhere, and in each according to its capacity; everything particular "comprehends in its mode the whole World Soul, although not totally as we have already pointed out; the World Soul is whole in any part whatever of the universe."[9] The World Soul is in each entirely, just as Substance is in each entirely. Everything we see of difference, transformation, figure, color, is nothing but a diversity of appearance; each thing is a mutable, transitory mode of what is immutable and eternal: "All that makes for diversity of genus, species, difference, and properties . . . all that consists in generation, corruption, alteration, and change, is not being or existence, but a condition and circumstance of being and existence, which is one, infinite, immobile, subject, matter, life, soul, true, and good."[10]

From the fact that "all contraries coincide in the One," it follows that: (1) all things are one; (2) every number, even or odd, is reduced to unity; and (3) since the principle of this coincidence is a principle of nature, "he who wishes to know the greatest secrets of nature should regard and contemplate the minimum and maximum of contraries and opposites."[11]

"De la causa" ends with these summaries. Taken together, they are the product of Bruno's attempt to determine the constitutive principles of the universe. Throughout the investigation, each step was directed by the part it played in the achievement of the final goal, which was that of attaining, through an application of the four causes to the universe as a whole, a definitive statement of those principles which are constitutive of that Nature. At no point in the discussion was there any attempt to apply these principles to the universe, in any other than metaphysical terms. As we shall see, this application is the goal of the "De l'infinito," the analysis of which is to be undertaken next. Before proceeding, therefore, to the second aspect of the problem of infinity, it will better serve our purpose at this time to place alongside the conclusions of Bruno, the formulations of Plotinus and Cusanus. By exhibiting the differences as well as the similarities between their formulations and that of Bruno, the doctrine of the latter may be clarified.

Plotinus accepts the view that the world is directed by a spiritual Principle, but for him, the Principle is not of the same nature as body; though the World Soul is incorporeal and entirely "present" in things, its unity is not dispersed throughout the things that it animates. The soul has no common nature with body; it has, consequently, no common measure with the body, for it is equally alike within the whole and within the part. "We are not to parcel it out among the entities of the multiple; on the contrary, we bring the distributed multiples to the unity. The unity has not gone forth to them; from their dispersion we are led to think of it as broken up to meet them, but this is to distribute the controller and container equally over the material handled."[12] For

Bruno, the World Soul is an active Principle; but it is not a Principle different from or inferior to the Universal Intellect.[13] For Plotinus, the First Principle of all things is no one of those things nor any sum of them; it remains simple, against the multiplicity of its "products."[14] The One, for Plotinus, is not intellect, for the intelligence is at one and the same time intelligence and the intelligible; the One cannot be either of these because they are not either; the One precedes this distinctive quality between intelligence and the intelligible. Nor can the One be all, for this would negate its nature as Principle:

> The universe springs from a source, and that source cannot be the All or anything belonging to the All.[15]
>
> Generative of all, the Unity is none of all; neither thing nor quantity, nor quality nor intellect, nor soul; not in motion, not at rest, not in place, not in time; it is self-defined, unique in form, or better, formless, existing before Form was, or Movement, or Rest, all of which are attachments of Being and make Being the manifold it is.[16]

Though Bruno admits that the One Substance is without form and determination, he will not admit that it is therefore transcendent, for Bruno identifies the One with the all; Substance is immanent in all, not only by presence of power, but also by actual presence in all.[17] For Plotinus, the One engenders the Intellect,[18] which in turn begets the world. The One, for Bruno, can engender nothing, since it is everything; the One, for Bruno does not "engender" the World Soul either immediately or through an intermediary. Nor does it "engender" the Intellect; both the Intellect and the World Soul are aspects of a Substance, for which the concept of generation can only mean generation of "modes." For Bruno, God is in all, and all is in God[19]: "What engenders and what is engendered is one."[20] The principle of resolution, for Bruno, remains the principle of *distinction*; the principle which governs Plotinus, in so far as his theory of hypostases is concerned, is the concept of *separation*.

Though Cusanus, whom Bruno follows closely, employs the same premises as does Bruno, he yet differs from the latter in his application of his principles. For though Cusanus holds the world to be in God, and God to be in the world, he bases himself on the principle of causality. Thus he declares that God is the complication of all, and that the world is the explication and development of God, only because God as Cause is in the world as a cause is in its effect, and the universe as effect is in God as an effect is in its cause. At the same time he holds the universe to be the finite image of the infinite God, while indicating that the universe can be called infinite only in the sense that it is the greatest of created things. The world, for Cusanus, is not eternal in the precise sense that God is eternal; in relation to God, the sensible world is finite. In spite of the fact, then, that the created world is a relative infinite, and in spite of the fact that the world is in God as one and infinite, Cusanus still declares himself for "dualism," insisting on the creation of the world out of nothing.[21] Bruno accepts only such conclusions as are consistent with his principles; contrary to Cusanus, he insists on the infinity of the universe, but because of the fact that he too applies the principle of causality: "An infinite cause must have an infinite effect."[22] Substance, in short, is unique and one wherever it is.[23]

"We must not believe, that there is in the world a plurality of substances, and a plurality of that which is truly being."[24] For Bruno a "finite substance" is a contradiction in terms; to be finite is to be limited, to be determined, to be bounded. By definition, then, it excludes existence "per se"; since there is but one Substance, it is unique and incomparable; and being unique and incomparable, it is as such inconceivable in itself; it can be conceived only through its accidents, and hence cannot be conceived at all. The "finites" are in a sense the vehicles of manifestation, and it is these that Substance "employs" to externalize its immanent activity. To be parts, portions, members, wholes, equals, greater, lesser, this, that, of this, of that, coordinated, different—these do not exist in and of themselves; they "are," through being referred to substance. The multiplicity of modes is but a multicplicity of "expressions"; Substance is in all things and all things are in Substance.[25]

Chapter V

CONCERNING THE INFINITE UNIVERSE AND WORLDS

In "De la causa," Bruno distinguished between the infinity of Substance and the infinity of the universe. He indicated that the Infinite Substance is all that can be, and contains in itself all being, "because it is all that any other thing is and can be, and can be what any other thing is or can be." He also said that the universe is all that can be, yet that no one of its "parts" is all that can be; the universe is all that can be in a developed, dispersed, and distinct way, while the Infinite Substance is all that can be, unifiedly and indifferently: "It is all, and as a whole, in each without distinction and differences."[1]

The "De l'infinito" deals with the infinity of the universe in order to apply to the universe and its worlds the conclusions gained in "De la causa." This will become evident in each particular resolution, where the problem is in each case treated under the heading of the infinity of the universe, while at the same time it is not forgotten that Substance is in the parts, and the "parts" are not "in" Substance.

"De l'infinito" attempts to refute the arguments presented by Aristotle in the *De coelo* and in the *Physics*; the greatest emphasis is placed upon those arguments in *Physics*, iii, 4. 5, and in *De coelo*, i. 5. 9.[2]

In "De la causa," Bruno had been satisfied to speak of the universe and its relationship to Substance from the point of view of the actuality exhibited by each. He had therefore made the distinction between that which is all, and has the capacity to be everything that everything else can be, and that which, while it is all, cannot be everything that everything else can be. This meant that Substance must actually be everything because potency and act, when absolutely considered, are identical. In "De l'infinito" Bruno carries this distinction one step further, in order to facilitate the formulation, by setting against one another the differences between the two "kinds" of Infinity.

> I call the universe all infinite because it has no margin, term, or surface; I do not call the universe totally infinite because each part that we encounter is finite; I call God all infinite because he excludes from himself every term, and every one of his attributes is one and infinite; and

> I call God totally infinite because all of him is in all the world and in each part of it infinitely and totally.[3]

The first kind Bruno calls "extensive infinity"; the second, he calls "intensive infinity."[4] The "De l'infinito" deals with the first, that is, with extensive infinity.

Bruno's treatment of the problem concerning the actual infinity of the universe accepts, as its point of departure, the Aristotelian disjunction, namely, that the world is either finite or infinite. Aristotle, according to Bruno, employed this disjunction in order to reach the most complete solution. Aristotle's intention was to prove the world to be finite by reducing the conception of the world's infinity to absurdity. Bruno accepts this method as his own, with the intention of proving the contrary; that is, he intends to prove that the conception of the world's finitude is absurd, thus leaving no alternative but the acceptance of its infinity.

"If the world is finite, and outside the world there is nothing," as Aristotle held, "where then is the world?" Aristotle had answered both questions with "in itself."[5] This answer, says Bruno, follows legitimately from Aristotle's definition of "place"; but is the definition correct? If it is not, neither the definition nor the conclusion derived from it is to remain a part of true knowledge. In stating this definition, Bruno declares, Aristotle has so conditioned the problem that the answer is one that is contained in the point of departure; by defining "place" as the "innermost motionless boundary of what contains,"[6] he has set the stage for his leap to universal place, the convex of the first heaven. Thus he could say that the first heaven, because it was not contained, could not possibly be in place; and since it was not in any place, it consequently had no place, and thus was "in itself."[7] From this it follows logically, Bruno admits, that "outside the world there is nothing." However, many puzzling questions arise from this view; if, for example, in order to escape a vacuum you assert that outside the world there is nothing corporeal, but yet continue to hold that there is an intellectual and divine being, a completely transcendent God who is the "place of the universe of things, the question immediately arises, how can an incorporeal intelligible thing without dimensions be the place of a dimensional thing? If you answer that it is to be conceived as its form, and in the same manner as the soul is conceived to comprise the body, you do not answer the question concerning the outside."[8]

The argument summarized here is a recapitulation of *Physics*, iv. 5, where Aristotle says, "but other things are in place indirectly, through something conjoined to them, as the soul and the heaven."[9] Thus the heaven is not in place except indirectly, like the soul, for its parts are in place[10];

> The upper part is moved in a circle, while the All is not anywhere; for what is somewhere is itself something, and there must be alongside it some other thing wherein it is, and which contains it; but alongside the All or the Whole, there is nothing outside the All, and for this reason, all things are in the heaven; for the heaven, we may say, is the All.[11]

From the double-edged implication of the same argument, we can see that Aristotle's point can be turned to Bruno's advantage; Aristotle, seemingly sensing the criticism, anticipates the question and asks, "where will this place, which is self-existent, be?"[12] If everything has a place, place too will have a place, and so on ad infinitum[13]—which would reduce this view to absurdity. The criticism is taken care of by defining place as he (Aristotle) has done: "Thus if we look at the matter inductively, we do not find anything to be in itself in any of the senses that have been distinguished"[14]; "and obviously, a thing cannot be in itself primarily."[15]

Bruno is aware of these conjectures, as is evident where he says,

> And if you attempt to excuse yourself, and parry the question, by maintaining that where there is nothing and no thing, there is also no place, other, or outside, this too will be very unsatisfactory, because these are words and pretexts which have no concrete significance . . . further, if you characterize the Divinity as that which fills the vacuum, and is its ground, and is the exterior form that bounds the world, you will be degrading the Divinity and the Universal Substance, because all that is said to bound, and limit, is either exterior form, or is the containing body.[16]

You contradict yourself, and reduce your position to the absurdity of employing terms with no foundation in reality. "The divine nature is no less nor in any other manner within the whole than without"[17]; "were Divinity that which bounds space, it would itself be space under another name."[18] Aristotle's definition, according to Bruno, leaves no alternative but to rule out universal place: "What does not fit the definition, what has no place, in virtue of this definition, is the prime and greatest place; the latter is what contains only, and is not contained"; for "if it is the surface of the containing body, and is not joined and continuous with the contained body, it is a place without an object in it, since it does not belong to the first heaven to be in place, except through its concave surface which touches the convexitude of the second body." For Bruno, then, the definition is confused and false.

What Bruno is trying to show here can only be brought to its full meaning if we consider, as he does, that there are two surfaces involved, the concave and the convex; for unless the concave of the first sphere is continuous with the convex of the second sphere, there must be between the two another place, which is not taken into account by Aristotle. This "intervening" place, according to Bruno, deserves, more than any other place, the name of "universal place."[19] There is, therefore, besides that "place" which the "Aristotelians" call "the surfaces of the concave," another "place" which is the "place" between the contained and the containing. Failure to take account of this "intervening place" is made the more serious by the statement that the containing body is incorporeal and the contained body corporeal; seek that "sole boundary" which is the "first," and you seek unto infinity, declares Bruno.[20]

In the opening statement Bruno said that "if you persist in this mode of argumentation, you cannot escape a vacuum,"[21] and it is to this side of the question that he now directs his attention. This vacuum has no outer boundary, and is therefore bounded on one side only, which is more difficult to imagine than an immense and infinite universe. But apart from this difficulty, Bruno continues, an examination of the terms will reveal an internal difficulty, for it is imperative to ask whether all space is full, since, according to Bruno, that vacuum which is indicated by the position of the "Peripatetics" is also able to receive body: "for if this world, which is in space, were not 'there' . . . this space would be no different from that (outside it)."[22]

For Aristotle there is no vacuum, in the sense of empty space, "outside" the universe, as Bruno's argument supposes. For Aristotle, the finitude of the universe demands that there be absolutely "no thing" outside it.[23] For Aristotle, there is likewise a contradiction in supposing the existence of a vacuum within the world, because this would make natural motion impossible; for, since the void has no differences in it, there could be no reason for motion. Bruno's argument is against a position which Aristotle does not even recognize; against the true Aristotelian statement, Bruno has nothing to offer. Bruno's position is the result of his own identification of Aristotle's "being in itself" with the conception of filling a vacuum; this is based on Bruno's conviction that "neither body nor space can be thought of, one apart from the other."[24] There is no reason, he holds, to affirm that one part of space is filled and the other is not; there is no reason why this space where our world is, should alone determine the category of that which is in space.

In this infinite space, there are infinite worlds, similar to this, and not different in kind from this.[25]

There is one matter, one power, one space, one efficient Cause, God and nature, everywhere equally and everywhere powerful. We insult the infinite Cause, when we say that it may be the cause of a finite effect; to the finite effect, it can have neither the name nor the relation of an efficient cause.[26]

Thus far the appeal has taken the form of presenting two alternatives. If one holds the conception of finitude, he must also hold the conception of the existence of a vacuum; that is, of a "nothing" outside the finite universe. Now if this is held, it must also be admitted that the outside "vacuum" has the capacity to "have" a world, just as this space, where our world is, has. If, however, one does not affirm the existence of the external vacuum, then space is full, and consequently he must also admit an infinite space with infinite worlds. Both arguments lead to a rejection of finitude and an acceptance of infinity; however, both arguments are based on the assumption that were this world not where it is, this space would not be different from that which is called a vacuum by the "Aristotelians." There is here the presupposition that "nothing" is equivalent to vacuum; the two are indifferently regarded as having the capacity to receive bodies.[27] This follows from Bruno's principle, developed in "De la causa," which states that in the absolute, potency and act are identical; consequently, if in the space "outside" the world there is a capacity to be, there will also be joined to it its proper act. For Bruno total absence of matter is a contradiction in terms; absence of matter, as we have seen, means an absence of Substance.

Aristotle would never agree that the removal in thought of the bodies of this world could be accepted as a proof that this space and that space "outside" the world are indifferently one. For Aristotle, where there is no body there is no space: "If then a body has another body outside it and containing it, it is in place, and if not, not."[28] In principle the arguments are alike, in the sense that the problem is formally the same one; but "de facto" the departure is obvious.

A summary of this section shows the following points to have been covered:

(1) Bruno first analyzed the Aristotelian definition of "place"; this was shown to have left out of consideration that ultimate or universal place, which is the place of this world.

(2) By assuming the Aristotelian conception, Bruno has attempted to derive its inconsistencies; thus the infinity of the universe has been shown to be the more acceptable thesis. This was accomplished by stressing the lack of difference, revealed by analysis, between the "nothing" which exists outside the world and the "place" of this world.

Chapter VI

INTENSIVE AND EXTENSIVE INFINITY

In continuing his refutation of the Aristotelian arguments concerning infinity,[1] Bruno centers his attack on the following points:

(1) Aristotle's definition of the infinite, as that which cannot be traversed, is subjected to critical analysis in order to show that the "finite body" which Aristotle assumes in his definition leaves out the consideration of an infinite body.

(2) In speaking of "parts," in regard to the Infinite, the distinction must be made between "parts of" and "parts in"; the first distinction does not apply to the Infinite; the second is nearer the truth, though it is still nearer the truth to speak of the Infinite in the "parts."

(3) Action and passion are not to be applied to the Infinite, because the latter is both agent and patient.

(4) In all of the foregoing, the conception of intensive infinity is not to be divorced from the treatment of the infinity of the universe.

In his attack on the definition of body, Bruno sets out to prove that Aristotle has assumed what he has not yet demonstrated; this is made evident where Bruno says that Aristotle has taken it as an established fact that everything has a form, and had thereby excluded the Infinite because it can have none. Because of this, Aristotle has been able to draw this conclusion validly from his initial assumption. In this case the starting point was the assumption of an absolute center and an absolute circumference; Bruno's attack is therefore aimed at these bases upon which Aristotle had built his conclusions.

Aristotle has maintained, says Bruno, that an actual "infinite" is impossible.[2] He has attempted to show by an "either or" proposition that if there were an infinite body it would be impossible to describe it, because any such attempt would end in contradiction; "every body is necessarily to be classed either as simple or composite,"[3] and since the disposition of the composite is involved in the simple, it will naturally follow from a proof that the simple body is finite, that the composite is also finite. The first example is that which deals with circular motion; if we should draw radii from the center of an infinite revolving body, the radii would be infinite, and so also the space between the radii; and this would render circular motion impossible. But circular motion of the heavens is presented to us as a fact; therefore, we must accept the fact that the body which moves circularly is finite in every respect. The same conclusion follows from the second of Aristotle's problems. Starting with the statement that upward and downward movements are contraries, and that they are movements to contrary places, Aristotle shows that if one is determinate the other is also determinate. This is then applied to center and circumference, to show that both the center and the circumference are finite and determinate, with the result that body is again found to be finite and not infinite; "neither motion toward or away from the center can be infinite."[4] A third argument, which proves the impossibility of an infinite simple body, follows from the impossibility of circular motion being completed if the heavens are infinite: for, if AB is an infinite diameter, and a perpendicular is drawn at E, the revolving line CD will never be able to cease cutting E; the position will always be something like CE, CD cutting E at F. The infinite line therefore refuses to complete the circle.[5] A completed circular motion

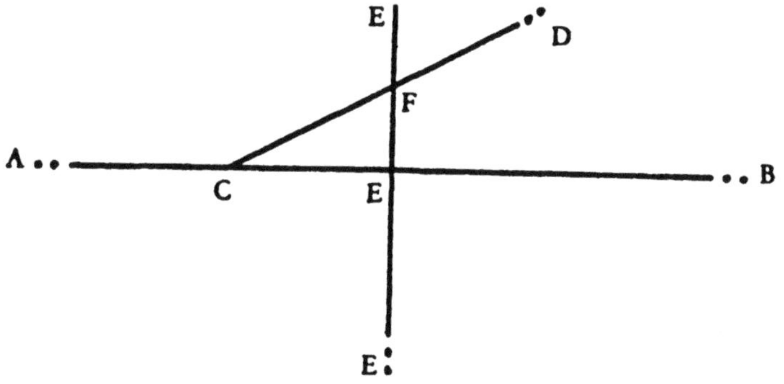

of an infinite body is therefore impossible. The result of these arguments is that several kinds of simple body have been examined, and no one of them has been found to be infinite.[6]

Aristotle, according to Bruno, has not treated the real nerve of the question. He is assuming many things in his report that do not really parallel the statements of the "ancients" on these points. Because the ancients are known to have declared themselves for an infinite body, it does not follow that they held that this infinite body had a center or an extremity; for the ancients, the infinite was a void, an ethereal infinite, and to this cannot be attributed the notions of gravity, levity, motion, region, or center. Centers are everywhere in the infinite, and nowhere in particular; centers are relative, and every world and every planet can be considered center from the point of view of one who is on that particular world or planet.[7] In the ethereal field there is no determined point to which heavy bodies move as to the center, and from which light bodies move, as Aristotle assumes. In the infinite universe, there is neither center nor circumference; in this infinite space there is a center everywhere and nowhere in particular, and a circumference everywhere and nowhere in particular. Nothing is heavy that is not at the same time light, because all parts change their positions successively; so that, according to its position as regards its movement on the earth from circumference to center and from center to circumference, a thing may be heavy from one position and may be light from another position. Things gather toward the center of gravity by force, and tend to diffuse by force of disintegration; nothing is heavy or light in its own position. Gravity and lightness are defined as attraction to a preserving position, and flight from the contrary.[8] In stating that the earth is center, says Bruno, Aristotle has committed a "petitio principii"; motion is not absolute, not from an absolute center to an absolute circumference; it is relative to position. In the infinite universe, motion is relative to the particular worlds within that universe. Absolutes of whatever kind are non-existent; of all the bodies that make up this infinite universe, none is heavy or light in itself; heaviness and lightness do not belong to the nature of a thing, but to the relative position which each particular thing has to another. These qualities are qualities of the "parts" in the universe; they are indistinct and indifferent from the point of view of infinity itself.[9]

In *De coelo*, i. 7, Aristotle returns to the problem concerning the actual existence of an infinite body. Where he had previously confined his investigation to kinds of simple body with a view to establishing their finitude, he now approaches the same question within a wider range, extending the data in a way which allows greater facility in application, and greater facility of demonstration. "Every body must necessarily be either finite or infinite, and if infinite, either of similar or dissimilar parts; if its parts are dissimilar, they must be of a finite or an infinite number of kinds. The kinds cannot be infinite . . . since the number of simple motions is finite, and, therefore, also the elements . . . If the kinds are finite, each of its parts must be infinite, such as, fire, water, etc.; this is impossible since it has already been shown that there is no infinite weight or lightness; it would also mean that the places of each are infinite in extent, and the movements of the bodies as well."[10] The argument can be summarized as follows: (1) neither that which moves downward, nor that which moves upward does so to infinity; (2) nothing in nature operates in vain, but moves toward an end; (3) since that is so, it is impossible that something should move toward that at which it can never arrive. Aristotle refuses to speak of a change, unless that change can be completed.

Aristotle's entire argument, for Bruno, is based upon an equivocation, involving the hidden assumption that one can proceed from finitude to infinity. "We must take, as our point of departure, the propositions of Democritus and Epicurus . . . there is an infinite plenum and vacuum, the one situated in the other . . . and there are diverse finite kinds, some included in others, and some relating to the other."[11] Though Bruno paraphrases the words of Democritus, Epicurus, and Lucretius, the principle of resolution that is brought to all problems, in this text and in the others which follow, is that there is but one Substance, identical wherever it is. In consequence of this principle, the "diverse kinds" have for Bruno, a meaning not altogether identical with the meaning of Democritus, Epicurus, and Lucretius.[12] Hence, for Bruno all these diverse finites concur in making up one Infinite Substance; "parts" are not only infinite in their total number, as they are for Lucretius, but also infinite in themselves, in the sense that the Infinite Substance is by definition "totally" in each one. Each of these "parts" is finite as presently conceived; but a broader view shows that in the Infinite, with its infinite transmutations, the "atoms" which constitute the universe in its physical nature,[13] will sometimes be presented to us as this particular, and sometimes as that particular; yet these "atoms" are one with the eternal Substance which never changes.[14]

To avoid misunderstanding, Bruno's position on "atomism" must be treated here in some detail. Ancient atomism, upon which Bruno based his conclusions, was materialistic. Though Bruno has admitted that he at one time also held the completely materialistic view, he has also stated his reasons for leaving that fold. While he retains some traces in his doctrine of the notion that physical operations are to be explained by the arrangement and movement of atoms, he also gives additional meaning to this statement by affirming that the "atom," as such, is no more than an initial base to be employed in the explanation of the structure of the universe as we experience it. Thus Bruno's meaning becomes understandable only in the larger context of what is called "intensive or substantial atomism," or even "metaphysical" atomism. From this point of view, the atom is the simple, indeterminate substance of all things. Since the universe which appears to sense is in constant flux, and since this flux is not thought to be the real, but to be the illusory appearance of that which is real, the conception of the atom is for us the result of our attempt to find the real Substance. This real Substance is itself unaffected by the change, which is a superficial thing. Number, plurality, and diversity are human determinations; they are not to be applied to Substance, because the latter is one and simple; the infinity of atoms is an infinity which is meaningful only when number, plurality, and diversity, are made the bases of cognition. For Bruno, then, the atom, is not only the atom of Democritus, the infinitely small particle, but it is at one and the same time soul and body, center of energy, and material subject, all of which are included in the operations and actions of a World Soul and World Intellect which encompass all and penetrate all.

A return to our text bears this out:

> Everything particular is in continuous alteration and change; that is, a change in the disposition of its atoms, and a change in place; the primal subject [the Infinite Substance] moves infinitely, that is, it does not move at all; the "parts" enter into and go out from this to that other place, part, and whole.[15]

In other words, Bruno does not oppose Aristotle on the point concerning the "completion of a change"; he maintains, however, that these changes do not affect the Infinite Substance. The argument is thus turned about, by Bruno's emphasis on the fact that Aristotle's entire discussion centers about the particular, whereas his own discussion is extended to the Infinite itself; consequently, to speak of a change with regard to the Infinite is nonsense. The attack is directed against another of Aristotle's arguments against the actual existence of an infinite body:

> Suppose the body to exist in dispersion; it may be maintained nonetheless that the total of all these scattered particles, say of fire, is infinite; but body we saw to be that which has extension every way; how then can there be several dissimilar elements, each infinite? Each would have to be infinitely extended in every way.[16]

Another statement of this problem would be: Since the number of places is finite, the number of elements must also be finite; and since it is impossible for each to be infinite in extent, therefore, the whole is necessarily finite.

For Bruno, this is still another result of the misunderstanding of the conception of the Infinite; one does not "add" all the dispersed particles —of fire, for example— in order to arrive at a totality which is infinite in its kind. This "totality" is not a natural one; it is no more than a logical, arithmetical, and geometrical weight and mass.[17] It does not at all follow that from this kind of conceptual addition one can reach an infinite body of one kind, but only that one can reach a kind of body in infinite finites[18]; these, though discrete, find themselves naturally and truly in one continuous 'infinite' which is the space, the place, and the dimension capable of containing all of them.[19] It is a contradiction to attempt to treat the actual Infinite as being composed of "parts"; this not only leads to the absurdity spoken of above, the belief in infinite weight and infinite lightness, but to the further contradiction of speaking of the place of the Infinite. The Infinite, Bruno points out, does not move; it is not mobile either in potency or in actuality.[20]

Turning to the problem of "interaction" between bodies, and its relation to the conception of infinity, Bruno restates the Aristotelian position that interaction between the infinite and the finite is impossible in a finite time. This follows from Aristotle's principle that the finite cannot act upon the infinite, and that the infinite cannot act upon the finite; for if this were possible, the finite would be admitted to be capable of performing that which is proper only to the infinite and the latter would thus become dispensable. The argument is stated in another way as follows: Since every sensible body is capable of acting or of being acted upon, it is impossible that an infinite body should be perceptible—for the infinite cannot be determined by a finite act; but since all bodies that occupy place are perceptible, it follows that no infinite body exists in the space outside the heavens.[21] For Bruno, there is never any such action and passion as Aristotle indicates; whether we view the case of the finite action between bodies, or an action between finites and infinites, or between two infinites, the agent never exerts its total vigor and power on the effect. Action and passion are extensive according to dimension and distance, and the agent and the patient are never so close that all their parts are continuous; consequently, the action will never be infinite, because the parts are not "intensive" but "extensive." This excludes the possibility of the Infinite working according to its total power in a part; its activity will be manifested discretely and separately according to part and part, since action in accordance with all the parts is impossible. Action and passion between two infinite bodies is an impossibility, since the one "possesses" the same power and resistance as the other; even if there could be such an interaction, no alteration could result from it. Intensive activity is conceded by Bruno, but the results are negative. The result of all action remains finite in effect: "The infinite (the whole) is immobile, unalterable, incorruptible . . . in it there are infinite and innumerable alterations, all perfect and complete."[22]

Chapter VII

THE INFINITE UNIVERSE

Bruno's arguments: *Motion, natural and unnatural, is to be considered in its own sphere. Distance between bodies is not sufficient ground to postulate differences in the nature of these bodies. Motion is determinate within finite worlds, but indeterminate in the immensity of infinity itself. Infinite velocity, weight, and lightness, do not exist in the nature of things.*

IN *De coelo*, i. 8, and *Physics*, iii. 5, which form the basis of the investigation in the third dialogue of "De l'infinito," Bruno considers the following points:

(1) From an analysis of motion, he finds that movement, natural and unnatural, is to be considered each within its own sphere; that is, each planet has its own intrinsic movement. Hence there is not only one center in the universe, as Aristotle held, but each planet has its own proper center.

(2) Distance between bodies is not a sufficient ground upon which to postulate difference in the natures of these bodies; the vital principle of a body has no intrinsic relationship to its accidents, such as its place and distance from other bodies; the vital principle of a particular body is one with the vital principle of the Infinite Substance, for the latter is totally in everything that exists.

(3) Each of the infinite worlds is finite as presented to us; but though motion is determinate within particular worlds, it cannot be inferred from this that motion is determinate in the infinite.

(4) Absolute infinites, such as infinite velocity, infinite weight, and infinite lightness, are not natural, but are logical, arithmetical, and geometrical conceptions; they do not, as such, exist in the nature of things.

For Aristotle, motion and rest are either natural or unnatural to bodies; each body moves naturally to the place in which it rests, and each body moves unnaturally to a place in which it "rests" unnaturally. From this it is inferred that since there is only one natural movement for each kind of element, it would be contradictory to assume that these other worlds, posited necessarily by those who affirm the existence of an actual infinite, consist of these same elements; for it would result that fire, for example, which has a natural tendency to rise, would be moving away from the center of one world towards the center of another world. Hence, either we do not admit that the elements are the same, in this and in the other worlds, or if we do admit this, it follows that there is but one center and one circumference, and that the world is one.[1] If there is such an actual infinite, its parts are either similar or dissimilar; and if similar, they should have the same natural motions; but if so, where will they move? Where, for example, would earth move, if the universe were nothing but earth? What inclination would it have for motion? If its natural place is infinite, then, wherever it is, it is in its natural place; or if it were supposed always to be in motion, it would then have no state of rest, a condition it must have in its natural place. In other words, if there are infinite natural places for the elements, there is in consequence no motion; but since motion is a fact, the infinite is not an actual existent. If the parts of the infinite body are unlike, then the natural places will differ; and since the number of kinds to be found among the parts will be finite or infinite, it is also necessary that if they are finite, some will be infinite in quantity and others finite. But this will naturally result in a condition where the infinite will overcome the finite; hence the infinite cannot have parts which vary in kind, if the number of kinds is finite. The only remaining alternative is that it have parts varying in kind, the number of which is infinite; but in this case the number of natural places is infinite, and so too, the number of elements. But it is impossible that the number of proper places be infinite; therefore, the whole must be finite.[2]

Bruno finds it impossible to reconcile himself to this reasoning. It does not at all follow, from the fact that all men possess the same physical makeup, that the parts of one move toward the parts of the other; there is no inclination of one part to move to the place of the other. In the same way, each world and each planet has its own parts and its own movement; the one does not move to the place of the other, nor do the parts of one move either naturally or unnaturally from one to the other; the motion of each is proper to each within its sphere.[3]

> As to us on earth, the earth appears to be the center of the universe, so to the inhabitants of the moon, the moon will appear as such. Distinctions of rising and falling are relative to the finite worlds, but they are not to be referred to the universe. Each world has its center, each its up and down; these differences are to be assigned relatively to the position of the parts of the whole.

It would be unnatural if the parts and the elements of our world should move towards the parts and elements of another, but not for the same reasons that Aristotle gives.

Throughout these arguments which Aristotle has presented in his refutation of the existence of an actual infinite, these important points stand out:

(1) The actual infinite cannot have infinites as parts; but the actual infinite cannot be without any parts.

(2) It follows that an infinite body cannot actually exist since (*a*) no simple body can be infinite, and therefore no compound body either; (*b*) different elements have different motions; and (*c*) movement and change is determinate because of these natural motions.

(3) Every argument presupposes these points, in one way or another; each discussion is merely another approach to the same problem. It is not surprising, therefore, to see Bruno repeating his answer in the majority of cases.

Is distance between bodies a sufficient basis upon which to postulate differences in their natures? For Aristotle this is unreasonable and may lead to the conclusion that all motion is due to constraint. "To postulate a difference of nature in simple bodies, according as they are more or less distant from their proper places, is unreasonable. For what difference does it make whether we say that a thing is this distance away or that."[4]

Aristotle's answer is not intended to be restricted to the particulars with which it deals; he is seeking to build upon this premise a conclusion which would hold that the heaven is one and can be no more than one.[5] The steps by which this is accomplished are: (1) "A body which has no movement at all cannot be moved by constraint"[6]; and (2) if bodies have a natural movement, the movement of particular instances of each form must necessarily have for its goal a place numerically one, that is, a particular center or a particular extremity[7]; and if it be suggested that the goal in each case is one in form but numerically more than one . . . we reply that the variety of the goal cannot be limited to this portion or that but must extend to all that have the same form[8]; hence, since only numerical difference can be postulated between homogenous portions of this world and those of another, and since a difference of goal finds its justification in a difference of body, there must be a different goal for every position of earth; this is impossible.[9] (3) The result must be one that abandons the assumption, or asserts the center and the circumference to be numerically one; but this implies that there is only one world, which was to be proved. The concept of "natural motions" is valid, therefore, and it is applicable to all the kinds of elements.

That the kind of motion which is natural to a body is not changed by greater or less distance from the goal is conceded by Bruno. But he denies that the first moving principle consists in the relation which a body has to a determinate place; rather, it is due to the identity of "the vital principle" and "the intrinsic desire" in the body, which is the evidence of its identity with the Vital Principle (Substance) which preserves itself in being. The "Vital Principle" is not essentially affected by the "accidents" of place, distance, etc.; we have an example, says Bruno, in weight and lightness, and in their relations to the particular worlds. Weight and lightness are not absolutes in the universe; they are not natural, but relative.[10]

It is with these notions that the fourth dialogue ends. Before proceeding to the treatment of the final dialogue of "De l'infinito," however, a word must be said on the notion of "the infinity of worlds" which has been employed by Bruno in this context to explain the "relativity" of motion and change.

To the intensive infinity of Substance and to the extensive infinity of the universe, Bruno seemingly adds a third type of infinity—the infinity of worlds. This "infinity," unlike the others, is one whose essential note is number, and must, therefore, be considered in its own context.

The infinite number of worlds is derived from the identity of power and will evident in the Infinite Substance; from the fact that the Infinite Substance comprises all particular things in itself, it comprises also innumerable worlds. In communicating itself to corporeal things, and in unfolding in particular existences, the reflection must be infinite in magnitude and number. That there are more worlds than this one is due to the presence everywhere of one eternal Substance; though it manifests itself in many ways, it embraces all plurality and number within itself.

> We are not compelled to define a number, we who say that there are an infinite number of worlds; "there" no distinction holds of odd or even, since these are differences of number and not of the innumerable. Nor can I think there have been philosophers, who, in positing several worlds, did not also posit them as infinite; for, would not reason, which demands something further beyond this sensible world, assume again and again, another and another, outside and beyond whatever number of worlds assumed.[11]

Chapter VIII

SUBSTANCE, UNIVERSE, AND INFINITY

Bruno's arguments: ***Time is common to our world and the infinity of other worlds. Infinite magnitude and infinite number coincide with infinite Unity. Bodies are not composed of simple elements, but of a mixture of many elements.***

THE FIFTH DIALOGUE of "De l'infinito" repeats many of the arguments of the preceding four dialogues; it lays particular stress on *De coelo*, i. 8, 9, and iii. 2. The following points are taken up for consideration:

(1) There is "time" outside of this world of ours; since time is the measure of motion, we can extend this notion to each world, which, because it moves by virtue of its intrinsic nature, has bodies which move similarly to those of this world; they can therefore be measured according to the notions of before and after.

(2) Infinite number excludes numerical degree and order; numerical degree and order depend upon the perceiver and the sphere within which he finds himself. If a transfer is made from the restricted sphere to that of the infinite, substantiality replaces relativity and infinite magnitude and infinite number coincide in the identity of one simple Being.

(3) Bodies are not composed of, nor constituted by, single elements; bodies are constituted by a mixture of elements, each body is named after the element which is predominant in each. Each element has something of the others in its structure.

If it can be shown that the universe contains in itself all matter, then the supposition that there either is, or can be, any body outside this world is a contradiction.[1] We recognize, says Aristotle, that in perceptible things, natural or artificial, we may distinguish between the form in itself, and the form joined to matter; we also recognize that any form has, or may have, more than one particular instance. On these grounds we might be led to infer that there might be more than one heaven; but "a thing whose substance resides in a substratum of matter can never come into being in the absence of all matter"; hence the universe can be shown to be one and not many, if it can be shown that this universe contains all matter. "If outside of this heaven, there is any body, either it will be simple or compound; and whichever alternative you take, I will ask further, whether it is in a natural location or in an accidental and forced location."[2] Since there are but three kinds of simple motion, and motion is related essentially to body, the two concepts of motion and body can be treated simultaneously; by proving that the universe has within itself all the simple and composite bodies, and with it their correlative motions, it can be shown that no one of these simple bodies exists outside the universe, because it is their nature to belong to this universe.[3]

Aristotle's argument appeals to the following facts:

(1) It is not possible for that body which naturally moves in a circle to change its place; for it is of the nature of circular motion that every point of its orbit is both a starting point and a goal.

(2) That body posited outside the universe cannot be the kind that moves from the center or that which comes to rest at the center, for these bodies could not naturally be "there" (outside the heavens), since their natural places are those that they now have. But if they could be "there" unnaturally, then these "exterior" places are natural to some other kind of body; but there is no other kind of body.

(3) The three kinds of simple body alone enter into the composition of this world, and they are moveable according to the three kinds of local motion; it follows, then, that outside this world there cannot exist any other simple body.[4]

(4) From this it can be inferred that there are not many worlds.

(5) It follows also that there is no space, either full or empty, outside of this world.

(6) Likewise it follows that there is no time outside this world because time is the measure of motion; and since there is no motion except that of bodies, there is no body where there is no motion; and where there is no motion there is no time.[5]

Bruno here bases his argument against Aristotle on his conception of a space which "intervenes" between every "Aristotelian" body and its containing limits. This intermediary place wherein every body can be moved is common to every kind of body or place that Aristotle names, and it is this which is the universal place or space of the universe: "There is no last end or margin of the universe; on the contrary, there is a universal space, in which there are other worlds like our own."[6] There may be, then, simple or composite bodies outside of this imagined extremity, "for just as the parts of this globe move . . . so may the parts of others move; and in fact, they move in no other way."[7] In infinite space, there is no reason and no necessity that can end or limit it; time exists outside this imagined circumference, for time is the measure of motions, and in the other worlds these depend on bodies similar to the ones moving in this world.[8]

The fifth dialogue ends with an argument which takes both Bruno and Aristotle out of the realm of the philosophy of nature into the realm of metaphysics. From the fact that circular motion was shown to be without beginning and end, and the further fact that it is one in cause and effect, Aristotle drew the conclusion that the unity of the universe which followed from these premises, was an example of the principle that a singular cause can have but a singular effect.[9] This was in turn applied to the first heaven, in order to show that that universe was one in rank as well as one in being. This was, indeed, a restatement in another way and from a different point of view of his distinction between the unmoved mover and the first moved, which forms the basis of the discussion in *Physics*, viii. 5 and *Metaphysics*, xii. In the former Aritsotle proceeds from two principles: (1) that everything which is moved is moved by another, and (2) that there cannot be an infinite series of movers and things moved. From these as premises, he concludes that there is a first unmoved mover, and also a first moved, which is not moved by itself, except in the sense that one of its parts moves another and thus moves the whole accidentally. The whole thing moved, is moved with an eternal motion, and therefore the power that moves it must be "infinite," and since "infinite," incorporeal.[10] Again, in the *Metaphysics* Aristotle indicates that into the discussion concerning the motion of things must also come the discussion of the desired good; for the desirable good moves without being itself moved. The first unmoved mover acts as the desirable good, while the first moved that moves itself is moved by the desire for the unmoved mover; and since the highest in the order of desires is the intellectual desire for the good itself, the first moved must have intelligence. Further, since only bodies can be moved, the first moved must be a body, but a body with an intellectual soul; this can only be the first heaven, which is one and infinite.[11]

Even if we were to admit, says Bruno, that there is a prime mover, it is not so entirely prime and principal that from it we can descend on a given scale to a second, third, and last mover. Infinite number does not allow for numerical degree and order; all these infinite spheres and movers have their true being and source in the one Infinite Principle and whole. All movers, all forces, can be reduced to one passive and to one active Principle, just as every number can be reduced to unity. Infinite number, infinite force, and infinite worlds, are to be treated in terms of the relationship and the proportion of the parts to the whole; for Bruno, whatever is an "element" of the Infinite must also be infinite. Thus we can have infinite earths, infinite suns, infinite worlds, and this does not mean that one infinite is greater than another; a thousand infinites are no greater than one, for the reason that the Infinite is not to be comprehended through numbers. In the Infinite there is not more or less, not a few or many; there is no measure or distinction which is real.[12]

A Summary of the Main Arguments Made in "De l'infinito"

Aristotle	Bruno
Aristotle's intention is to prove the finitude of the world by reducing the conception of actual infinity to absurdity. The definition of place, says Aristotle, makes it evident that the world is in itself," and outside the world there is nothing. The finitude of the world excludes the possibility of a vacuum; there can be "nothing" outside the world.	Aristotle's definition of place is sufficient only for "particular place," and leaves out any conception of universal place; there is no real difference between the "nothing" that exists outside the world and the "place" of this finite world.
Aristotle affirms that motion is from contrary to contrary, and the fact that one contrary is determinate necessitates that the "end" be also determinate; this fact excludes any motion to infinity.	This conclusion is based upon a *petitio principii*; it is founded upon a definition of the 'infinite' which by implication rules out infinity from the very start; for implied in Aristotle's definition is the "fact" of a determinate center, an assumption which is relative to finitude, and has no meaning for infinity.
Aristotle asserts that radii drawn from the center of an infinite revolving body will make circular movement impossible.	The assumption that an infinite body has a "center" from which radii could be drawn is itself false, and leaves Aristotle's entire formulation without foundation.
Neither that which moves downward nor that which moves upward can do so to infinity. If the infinite exists, then its parts are	The infinite "parts" that Aristotle speaks of are not infinite in themselves, but infinite only from the perspective of infinity. Human

necessarily infinite—a manifest contradiction.	conception is responsible for the appearance of parts; for it naturally conceives the infinite not as it exists in itself, but as it accidentally appears in particulars and as thus affected by their place and atomic composition.
Since corporeal substance, which has extension in every direction, requires more than one kind of element, each of these would be infinite.	In nature there are no separate infinites; all things concur in one Infinite, which is all things, and no one of them.
If the infinite moves, there must be "another" infinite place to which it moves, and therefore there is of necessity more than one "infinite" place.	The Infinite is immobile; motion is relative to the disposition of the atoms "in" the immobile Infinite.
Interaction between a finite and an infinite in a finite time is impossible.	Interaction has no meaning for an infinity which has no parts—each "part" that Aristotle speaks of has in it the "total infinite" which is one everywhere.
There is one "natural" movement for each kind of element; if, then, the same kinds of elements are "in" the other worlds which infinity posits, a contradictory situation would arise, wherein each element would be moving in a natural and unnatural direction simultaneously.	The argument is based upon the false assumption that there is but one determinate center in the universe. On the contrary, centers are everywhere; bodies maintain themselves by the transmutations of the infinity of atoms in the infinity of space.
If motion is infinite, so is velocity, weight, and lightness.	Infinite velocity, weight, and lightness, are abstract conceptions and have no foundations in nature.
From the fact that this world contains within itself all bodies, both	Substance which is the principle of nature is indifferent. There is

simple and complete, it follows that outside this world there can be no motion or time, because there is "there" no body; for time is the measure of motion, and where there is no body, there is no motion.

no "outside" such as Aristotle asserts.

Unity of the first mover implies unity of the universe—a single cause can have but a single effect.

There is no first mover such as Aristotle asserts; there is no effect transcendentally distinct from the First Cause. There is a First Principle and a First Cause, both transcendent and immanent, to which all movers are equally proximate and from which they are equally distant.

Chapter IX

CONCLUSION—THE CONCEPT OF INFINITY

WHATEVER REALLY IS, is Substance. The Substance of all is eternal, and only the outward form of it changes. Where things appear as parts, or as evil, or as imperfect, it is because we look at the part or the moment, and not the whole or the immutable. This is the lesson of the "De la causa" and of the "De l'infinito."

It is in the contemplation of the Infinite, that man attains his ultimate good; since all things strive toward the end which is intended for them by nature, the more perfect the nature, the more perfect is the tendency to fulfillment. The final goal for the individual human being is not to be found in a particular good or a particular truth, for these lead the individual from one thing to another; particular truths and particular goods serve only to show that there is more truth, and more good, to be known and to be desired. In each individual human being there is a desire to become all things; for this reason, each is directed to the Infinite, which is at once its cause, source, and end. The fact that there is this desire, and the fact that there is this quest for knowledge, makes mandatory the existence of that which can satisfy both, the Infinite Substance. This is the lesson of the "De la causa" and the "De l'infinito."

There waits for each being eternity and realization; the contemplation of the universe is the means by which the individual rises to the contemplation of the Infinite. The goal of each soul is to become united with its eternal source; that is, to escape from change and relativity, to truth, eternity, and immutability. This is not only stressed in the "Eroici furori," but also implied in the "De la causa" and the "De l'infinito."

The search for the Infinite is from the finite and measured to the unlimited and immeasurable; from the contracted "this" to the Substance which is all. Man cannot rest with that which is fleeting, or divided; he looks for that which is perfect, lasting, universal, and necessary. This is the lesson of the "De la causa" and the "De l'infinito."

We have already seen the method which Bruno employed in the pursuit of these goals. This method was in itself a result of the difficulties encountered in the pursuit; where Bruno seemed to be pursuing the paradoxical, he was in reality accepting the paradoxes in order to formulate his position. It remains now to summarize this position. Since the notion of "intensive" infinity is one that includes all, it takes precedence in treatment as well as in rank.

The characteristics of Infinity are:

(1) It does not move itself locally, because it has nothing outside of itself to which to transport itself—since it is itself all.[1]

(2) It does not generate itself, because there is no other being which it might desire or look for—since it is that which has all being.[2]

(3) It is incorruptible, because there is no other thing into which it could change—since it is that which has everything in itself.[3]

(4) It cannot diminish or grow, since it has no proportional parts; and consequently, it is that to which nothing can be added, and it is that from which nothing can be subtracted.[4]

(5) It is not changeable into any other disposition, because there is nothing external to it through which it might be passive, and through which it could be affected.[5]

(6) It comprehends in itself all contrariety, in unity and harmony; and since it has no inclination toward any other and any new being, or even to any mode of being, it cannot be the subject of change according to some quality; in it everything coincides.[6]

(7) It is not matter, because it is not configurated or figurable; it is not limited nor limitable; it is not form, because it does not inform or figure anything else; it is that which is all-one, greatest, and universal.[7]

(8) It is not measured or measurable.[8]

(9) It does not include itself because it is not greater than itself; it is not included by itself because it is not less than itself.[9]

(10) It is not equal to itself because it is not other and other, but one and the same; being one and the same, it has not being and another being; and because it has not being and another being, it has no parts; and having no parts, it is not composite.[10]

(11) In it the divisible is not different from the indivisible, the simplest from the Infinite, the center from the circumference; the Infinite is all that can be; it is immobile; in it everything is indifferent, and everything is one. It is all center, for the center is everywhere; in it the circumference is throughout all; consequently the center does not differ from the circumference.[11]

(12) It inhabits all the parts of the universe; it is the center of all that has being; it is one in all, and that through which one is all. Because it is all things, and because it comprises in itself all being, it brings it about that everything exists in everything.[12]

The notion of infinity is contained in each of these characteristics; this follows not only from analysis, but also from an application of the principle which each characteristic represents. The attempt here, as it has been throughout, is not to exclude anything, but to include everything. Hence the Infinite is not only extensive, but is extensive in a certain way; it is extensive in the same way that it has been called form, matter, soul, intellect, etc. The goal is to reach that conception which includes all in such a way that it transcends all things through not being exhausted by any one of them. The category of extension, for example, is one that possesses the notes of quantity, limitation, and measurement; it is the nature of infinity both to include and to transcend these notions of quantity, measurement, and limitation, for they are meaningful only when applied to the finite. No terminology such as "greater," "smaller," "equal," "unequal," "number," etc., can express an essential relationship to infinity. It is only when we prescind from extension, quantity, measurement, and limitation, and when we do not refer to the comparison that can be made between part and whole, that we approach the proper conception of infinity. When we pass beyond these limitations and transcend them, we arrive at that notion of infinity which has all limits, all terms, all determinations, and all ends, indifferently within itself; here, it is proper to speak of "the Infinite in the parts," rather than parts in the Infinite.[13]

Bruno, no less than anyone else, realized the difficulty of these concepts; it is for this reason that he, like Plato in the "image of the cave," sought to clarify these notions by means of image patterns. These are his "proofs" or "verifications"; they serve as aids to the understanding, and were not intended to be taken as actualities.

"What thing is more contrary to the straight line than the curve? What difference will you find between the smallest arc and the smallest chord? What difference will you find between the infinite circle and the

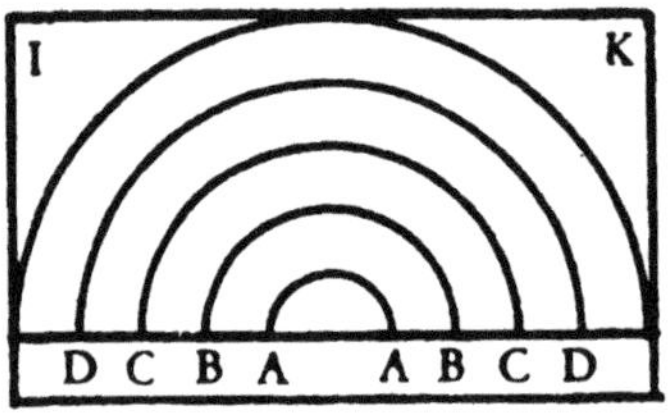

Arc BB is greater than arc AA
Arc CC is greater than arc BB
Arc DD is greater than arc CC

straight line?[14] "Therefore, the greatest arc becomes one with the infinite like IK at its point of union with it."

In other words, all limits are withdrawn at the point of union; they are "neglected" in the indetermination of the line without limits. Since there are no limits, there is no basis for comparison; the "note" of incomparability is one of the notes of infinity.

The conclusion, that all limits, ends, and determinations, are indifferently "in" the Infinite, is "verified" in the following manner: "If we take an infinite line BD to which is appended a perpendicular CM, we can show that as this perpendicular moves towards the line BD, all the angles that it makes with the latter, whether acute, or obtuse, whether greater or smaller, are one in the infinite principle, line BD."[15]

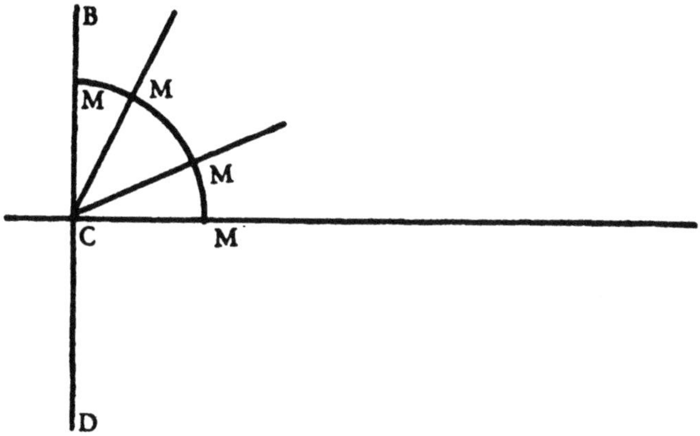

One angle can only be compared to another, as greater or lesser, equal or unequal, when the basis of conception is that of quantity or extension; but when the limits of all the angles find themselves indeterminately in the infinite angle, the concept becomes one of "intensive" infinity, and thus greater, smaller, equal, and unequal, become fused in the one that is all indifferently.

These conclusions aid in the interpretation of those answers given to the Aristotelian arguments; this is clearly seen if we recall several from the summary given in chapter eight:

(1) In the Infinite there is no center and no circumference.

(2) Therefore, we cannot argue as Aristotle did concerning circular motion, assuming a fixed center and a fixed circumference.

(3) There are no parts in the Infinite, for parts are relative to the perceiver, and bear no relation to the Infinite, since there is no relationship between part and whole.

(4) In nature there are no separate infinities; all things concur in one Infinite, which is all and none.

The full significance of Bruno's definition can now be seen: "I call God all infinite, because he excludes from himself every term, and every one of his attributes is one and infinite; I call God totally infinite, because all of him is in all the world, and in each part infinitely and totally."[16]

Though it has been shown that Bruno distinguished between intensive and extensive infinity, it has also been shown that this remains a distinction and it is not to be confused with a separation, nor with a real distinction of principles. This concept has been treated at length throughout the preceding chapters, and it need not be "labored" here. Indeed, if one were to seek a description of the philosophy of Bruno in the manner in which it is presented, the most suitable would be "The Philosophy of Identity."[17] Hence:

Intensive infinity is such that it is all that can be, and such that each one of its parts is all that can be. It includes as one aspect that infinity which is called extensive infinity; for,

The universe is infinite in extent, infinite in the number of its parts, and is constituted by elements which display uniformity of Substance. It is the omniform image of the omniform Substance; it is the reflection of Substance; it remains distinct only in aspect, and not in reality.

There remains now the consideration of the various applications of the concept of infinity; this will entail and examination of:

(1) Change
(2) Motion
(3) The coincidence of opposites and contraries
(4) The attributes of Substance

Change is not a change of being, wherein some particular thing seeks or achieves new being; change consists in the change of the modes of the one Infinite Being, which remains eternally identical with itself. The difference between the universe and the particulars in that universe is that the former comprises in itself all being, and all the modes of being, while each particular has not all the modes of being, though as Substance it has all being, in the sense that Substance is one wherever it is.

> Whence that (the Infinite) comprises all totally, because outside and beyond it (the Infinite Being) there exists nothing; that is, because it has no outside and beyond; of these (particular things) each comprises all being, but not totally because beyond each there are infinite others. Therefore, it is to be understood that all is in all, but not totally and in all modes in each one. Everything is one, but not in the same mode.[18]

Entity, substance, and being are therefore one being. From the fact that in the Infinite there is found multiplicity and number—a fact which has its cause in the existence of the modes and the accidents of being, each of which is called a particular thing—there does not result the multiplication of being as being; for "all that makes difference and number, is pure accident, pure figure, and pure complexion. Every production, of whatever kind it may be, is an alteration, with the Substance always remaining the same."[19]

> That which is in the universe, in relation to the universe, is throughout all, according to the mode of its capacity, in whatever relation it may be to the other particular bodies; because it is above, below, innermost, right, left, and according to all local differences; and because in the infinite, there are all these differences, and no one of them.[20]

There is in the world no plurality of substances, and consequently, no change from one substance to the other:

> Everything we see of difference of bodies . . . is nothing other than a diversity of appearance, a transitory, mobile, corruptible appearance of an immobile, stable, and eternal thing; in which all forms, all figures, and numbers are indistinct and as it were conglomerate, not otherwise than in the seed, in which the arm is not distinct from the hand, the breast from the head, the nerve from the bone.[21]

Quantity and measure are not Substance, but about Substance; Substance is essentially one and indivisible, and therefore without number and without measure. It is the subject of particularity that has measure and number: "Certain accidents of subsistence cause "multiplication" of Substance. In such a manner, certain accidents of existence cause multiplicity of entity, truth, being, and the one."[22]

Motion too, for Bruno, is not absolute, but a movement of the accidents of being. Motion is an aspect of the "change" which takes place between the modes of being. This is shown in the following way: If change is a change of mode and not a change of Substance, then there is no real significance attached to the distinction between immobility and mobility. There are two principles of motion, which in nature can be reduced to one root: (*a*) finite motion, which moves in time, with respect to the finite subject; and (*b*) infinite motion, which moves instantaneously, with respect to itself; those bodies that are moved by virtue of the Infinite, are not moved at all, because instantaneous movement and immobility are identical.[23] The "verification" of the identity of motion and immobility is given in the image pattern of the sun's motion: "Here is the sun; the sun is not all that can be; the sun is not in all the places where it can be; when it is in the east, it is not in the west, nor in the north, nor in the south, nor in any other position."[24] This is finite motion in finite time; but,

> When we want to show the manner in which God is the sun, we shall say (because He is all that can be) that He is simultaneously in the east, west, south, and north, and in whatsoever part of the earth's globe . . . and, therefore, if He is all that can be, and if He possesses all that He is able to possess, He will simultaneously be throughout all and in all; He is the most moveable, and the fastest, in such a way that He is also the most stable and immobile.[25]

The multiplicity of movers and of things moved, like the multiplicity of particulars, coincides in one simple Substance.

To the "modes" of motion and change Bruno adds the third "verification," the principle of the coincidence of opposites. No contrary exists except in relation to its opposite; the opposite is its limit and term, and thus its second or negative aspect.[26] Thus the first contrary is its end and has also a relation to the opposite and exists in the opposite as its beginning.[27] At the point of union where the term and limit of the one opposite fuses with the beginning or principle of the contrary which limits it, they are one; this relationship is characteristic of "finite" or "determined" being.[28] This determined "finite" therefore involves two aspects within itself; it contains being in itself, as the first contrary, but as involved with its opposite, the second contrary. Since it contains this notion of a limiting contrary, the first contrary can never be determined or finite or limited without its other correlative aspect, the second contrary. Hence, its existence as determined, finite, and limited by its contrary, cannot "be" unless it is so determined. Through this limitation, this determined something becomes a finite thing, since it is obvious that where the opposite "began," it, the contrary, is not, but has its determinate end. Only by its finitude, then, as determined, as limited, as contrary plus its opposite, is the first contrary or opposite a determinate "this," and thus a particular finite. This then is the very notion of finitude: that the opposite, the limit, the end, which by its nature of limitation "determines" the first contrary to be a finite this, is that in which the limiting contrary is not, since it is the principle of the other, and since it is the principle of the other, is contrary and opposite at the point of union. The determined or limited, then, in one sense coincides with its opposite, and in a sense does not. Only thus can a "this" be determined as a particular—that is, when it is limited by its contrary; and only "as opposite" is it therefore something determined, limited, and finite.[29]

Since a finite is a finite through union with its limiting opposite at their point of union, the Infinite must be that which can and does negate determination and limitation itself, and by so doing returns into itself indifferently and indeterminately all that is different and determinate.[30] In this notion of infinity the negation or opposition has not disappeared completely, but has been incorporated within the Infinite and transcended by it.[31] Infinity as such, therefore, would in a sense become one with all it comprehends in itself.[32] The Infinite would thus be that which is its own "opposition," its own "determination," its own "end," and that which by its own nature is a "negation" of any and all determinations—it would be that which "determines" and "limits" itself. It would thus be self-determined, and would be "in" itself. This Infinite which "determines" itself must be self-sufficient, then, by definition, and at one and the same time have within itself all the "limits" and "opposites" which would "exist" in it as these finite determinations—determinations which are but particular participating contractions of the infinite essence.[33]

It is the Infinite that can by determining itself return to itself, and while remaining one and self-sufficient reflect its particular participations. Again, by this inner negation of all particulars it necessarily becomes an affirmation of all of them, because the particular participating limitations have coincided in the Infinite, and have an existence as "related" to it.[34]

For Bruno, the "approach" to the secrets of nature, and the knowledge of the divine Infinite Being, lies in the true understanding of the "secret" contained in the coincidence of opposites and contraries. To know that there, at the point of union between opposites and contraries, we have a "both-and" and a "neither-nor," to know that the limit for one is a principle for another, is to be led to an approximation and higher understanding of the self-limiting, self-determining "end," which is at once an end without an end. From this we can draw three major conclusions. First, the Infinite is to be conceived as transcendent of particular determinations and limitations. Secondly, the Infinite is not "other" or "opposite," though it includes all particular others and opposites in that it is itself "particular" when predicated of finite things; the Infinite thus has within itself all that appears as multiple manifestations, since manifestation is but an unfolding and development of what is enfolded and enveloped. Thirdly, the Infinite must "transcend" itself, in the sense of producing its particular contracted modes; but these particular modes are no other than the Infinite itself as contracted to this or that particular determinate being.

We are thus in a position to understand, from another point of view, the distinction that Bruno makes in the "De l'infinito" between intensive and extensive infinity.[35] The intensive Infinite excludes all determinations, limitations, and restrictions, in the precise sense indicated above; that is, it is transcendent of particulars and includes them all. The extensive Infinite also excludes all determinations, limitations, ends, and restrictions not intensively or metaphysically, but extensively or physically; it is presented to us as finite because of its part by part apprehension. It is, so to speak, "finitely" infinite, while the intensive Infinite is infinitely infinite. Like the other distinctions that have been made between form and matter, potency and act, intellect and being, intellect and world soul, these two distinctions remain conceptual and not real distinctions, and depend upon our ways of conceiving the Infinite; for in reality they are one.

This conclusion receives further corroboration from a treatment of the Universal Intellect as an aspect of the Infinite. For Bruno the Universal Intellect is "the most intimate, the most real, and the principal faculty and effective partial power of the World Soul."[36] This intellect, through its function as efficient Cause,[37] infuses and brings something of its own into matter, thus producing everything that exists. Since it is a particular aspect of the World Soul,[38] and is distinguished only logically from that soul, it assumes its character as a Principle which comprehends in itself all reality; it is thus Substance itself, as intelligent. It is an intelligence which thinks things and which is, at one and the same time, all the things that it thinks. Thus though it has one of the characteristics of the Aristotelian intellect,[39] in that its own thinking is its object, it yet differs from the latter in that it is not set off against a world which it transcends. Bruno's Universal Intellect encompasses all that exists.[40] It is at once that which thinks and that which is understood. It is thus that Bruno says, "The divine mind and the absolute unity, without any species at all, is itself that which understands and that which is understood."[41] Truth is thus, for Bruno, itself being, necessity, principle, end, and perfection of all that exists.[42] Truth is before all, as Cause and Principle; it is in all as Substance. Truth which is in the Universal Intellect as constitutive of things is before things as their efficient Cause; it is in things as Principle; it is after things as the adequation of the thing with the particular intellect (the Universal Intellect contracted to be) which knows it as object. It is thus that Bruno says, "I want you to note that there is one and the same scale through which nature descends to the production of things, and the intellect ascends to the cognition of them."[43] Particular intellects thus manifest the simplicity of that intellect of which they are but particular modes,[44] for

> The intellect, wishing to free itself and loosen itself from the imagination to which it is joined, besides recurring to mathematical and imaginable figures in order that it may, either through them or through their similarity, understand the being and substance of things, also comes to refer the multitude and diversity of species to one and the same root.[45]

Again, "When the intellect wishes to comprehend the essence of a thing, it simplifies as much as possible; it retires from composition and multiplicity, by discarding the corruptible accidents, the dimensions, the signs, and the figures, from that which lies under these things."[46] "In rising to perfect cognition, we proceed by simplifying the manifold; just as in descending to the production of things, unity proceeds by unfolding itself. The descent is from the one being to infinite individuals and innumerable species, the ascent is from them to being."[47]

The distinction, then, between intellect and being, like the distinctions of form and matter, God and nature, is not a real one.[48] As has been pointed out above, the distinction between the Universal Intellect and the World Soul, which as Universal Form is the Active Principle of things, is one of function.[49] Thus, since the universal form works through the universal intellect from within matter to produce all that exists, we can arrive at the following formula:

(1) Universal Form, through its principal faculty, the Universal Intellect, *is the* Active Principle of things
(2) But the Universal Form is only logically distinct from the Passive Principle, Matter, so that
(3) The Universal Intellect, the Universal Form and that through which it works, Matter, are identical as Universal Substance.

Bruno's concept of infinity helps to give a tentative answer to a question which has been debated among his interpreters—that is, is the philosophical system of Bruno pantheistic? If what we have said is correct, there remains no difficulty in classifying the system as such, especially if it is granted that the characteristic note of pantheism is the identification of God and Nature. Our solution rests with the assertion that it has been clearly shown that all distinctions in Bruno's philosophy remain distinctions and not separations; it has also been shown that these distinctions are not distinctions of principles, such as the distinction between existence and essence, but that they are logical distinctions of what in reality is one and simple. If it is argued that even those who hold God to be completely transcendent also hold that the simplicity of God's nature is not contravened by the distinction of His attributes, it can be said in reply that the problem is not one where we admit that God is transcendent, but one that deals with the application of the concept of distinction to a Substance which is one and simple wherever it is. The quarrel is not one between separation and distinction, but one between distinctions which are real and distinctions which are logical, as they are applied to a system wherein every endeavor is made to show that God, Nature, Intellect, Form, Matter and Soul, are one. As to the further question, whether this pantheism is personal or impersonal, we believe that Bruno's answer would be: I have agreed that intelligence is the principle faculty of the World Soul; I have also agreed to the fact that this intelligence and this World Soul are aspects of Substance. If you ask me now, whether God or Substance is personal, I would have to say, yes and no. Substance is personal, but not only so, for Substance is not to be characterized by "one" of its aspects; it can only be described by these aspects, with none of which it is absolutely identified. Substance, in short, is like a one-way street; that is, whatever I can say of it is representative of it, but I cannot say that "It" is completely represented by anyone of the aspects nor by any addition of these aspects. The picture of Substance, the universe, looks like Substance, but I cannot say that Substance looks like the universe.

Giordano Bruno
of Nola

CONCERNING THE CAUSE, PRINCIPLE, AND ONE

To the most illustrious
Sir De Mauvissière

Published in Venice
in the year 1584

INTRODUCTORY EPISTLE

Addressed to the Most Illustrious
Sir Michael De Castelnau
Sir De Mauvissiere, Concressault, and Joinville,
Knight of the Order of the Most Christian King, Member of
His Private Council, Captain of 50 horsemen
and Ambassador to Her Majesty, Queen of England

Most illustrious and honored sir: if I again divert the eyes of my thought to your patience, perseverance, and solicitude, with which—by adding favor to favor and benefit to benefit—you have bound, obliged, and attached me, and are used to overcoming every difficulty, liberating me from whatever danger, and bringing to an end all your very honorable intentions, I come to discover how appropriately the noble device, with which you adorn your terrible helmet, fits you; (I refer to) that liquid humor which wounds suavely; so much so, that while it runs continuously and frequently, it softens, splits, subdues, cuts, and levels—by force of perseverance—a compact, rough, hard, and uneven rock.

When, on the other hand, I recall how—even leaving aside the rest of your honorable deeds—through divine ordination and high providence and predestination, you are my efficient and solid defender against the unjust outrages which I suffer—which are such that a truly heroic spirit was necessary in order not to abandon the work, nor to despair, nor to give up as conquered to such a rapid torrent of deceitful crimes, with which they had attacked me with all the power at their command: the envy of the ignorant, the presumption of the sophists, the censure of the malicious, the murmuring of the slaves, the whispers of the mercenary, the objections of the servants, the suspicion of the stupid, the apprehension of the informers, the zeal of the hypocrites, the hatred of the barbarous, the fury of the vulgar, the lamentations of those that I have resisted, and the cries of those whom I castigated; in all, the only thing that was missing was a low, insane, and malicious rancor of a female, whose false tears are often more powerful than the most arrogant waves and violent tempests of presumption, envy, detractions, murmurings, treasons, ragings, irritations, hatreds, and violences—I see you as that solid, firm, and inviolable rock, which emerges again and again, and shows its crest above the agitated sea; and is neither moved nor shaken, neither by a menacing sky, nor by the horror of the winter, nor by the violent jerkings of the agitated waves, nor by the confused torments of the air, nor by the strong blowing of the north wind; on the contrary, you become fresh again, and you become more invested with a similar rock like substance. You are therefore, endowed with a double virtue—through which the liquid and pleasant drops become most powerful, and the violent and tempestuous waves become vain; through whose work against the drops, the violent rock becomes so bland; and against the waves, the battered cliff arises so powerful. You are that selfsame one, who is the secure and tranquil haven for the true muses; and the ruinous rock against which the deceitful volleys of the impetuous plannings of the hostile ships are dashed to nothingness. I, then,—whom no one could ever accuse as being ungrateful, whom no one could ever censure as being discourteous, and of whom no one can justly complain; I, hated by the fools, depreciated by the vile, censured by the ignoble, dishonored by the wicked, and persecuted by the beasts; I, loved by the wise, admired by the learned, celebrated by the great, esteemed by the powerful, and favored by the gods; I, through so many favors from you—have been received, nourished, defended, freed, retained in safety, maintained in port, and rescued by you from great danger and tempest; I consecrate to you this anchor, these ropes, these broken sails, and these pieces of merchandise, which are very dear to me and very precious for the future world; so that, with your favor, they shall not become submerged in the unjust, turbulent, and hostile ocean. When thus they hang in the sacred temple of Fame, they will become powerful against the audacity of ignorance and the voracity of time; and they will also give eternal testimony of your unconquerable favor; so that, the world will know that this noble and divine product—inspired by great intelligence, conceived with tempered sense, and brought to light by the Nolan Muse—has not, thanks to you, died in its swaddling clothes, but has been promised a long life, that is, as long as this earth, with its living back, goes on revolving in the eternal sight of the rest of the shining stars. I have here that philosophy in which there is contained, truly and certainly, all that which, in contrary philosophies, has been vainly sought. First, however, let me offer you, in a brief summary of the five dialogues, an account of all that seems to concern the true contemplation of the Cause, Principle, and One.

Argument of the First Dialogue

In the first dialogue, you have an apology, or something else I know not what, concerning the five dialogues of "Le cena de le ceneri."

Argument of the Second Dialogue

In the second dialogue, you have first, the reason for the difficulty of that kind of knowledge. This is emphasized in order to understand how much the knowable object is removed from the cognitive faculty. Second, in what manner and to what extent, things caused and things proceeding from a principle, come to clarify the Principle and Cause. Third, how much the knowledge of the substance of the universe, contributes to the knowledge of that on which the universe depends. Fourth, by what means and by what road, we, in particular, try to know the First Principle. Fifth, the difference and the concordance, the identity and diversity, between the meanings of the terms: Cause and Principle. Sixth, what that Cause is, which is divided into efficient, formal, and final; in how many ways the efficient Cause is designated, and in how many ways it is conceived. In what manner the efficient Cause is, in some way, intrinsic to natural things—through being nature itself—and in what manner it is, in some way, extrinsic to those things. How the formal Cause is united to the efficient Cause, and is that through which the efficient Cause acts; and how the same is taken from the bosom of matter by the efficient Cause; how the formal and efficient Causes coincide in one subject and principle, and how the one Cause differs from the other. Seventh, the difference between the formal universal Cause—which is a soul through which the infinite universe, as infinite, is animated, not positively but negatively—and the formal singular cause, multipliable and multiplied to infinity; which, in so far as it is in a subject which is more general and superior, is so much more perfect; whence the great animals, like the stars, must be esteemed to be divine beyond comparison, that is, most intelligent without error, and operating without defect. Eighth, how the first and principal natural form, the formal principle and efficient nature, is the soul of the universe. This soul is the principle of life, vegetation, and sensibility in all things that live, vegetate, and feel; and it is held, by way of conclusion, that it is something unworthy of a rational subject to believe that the universe, alike with all the other principal bodies are inanimate, since it is a fact that from the parts and residuals of those are derived the animals which we call most perfect. Ninth, how there isn't anything, no matter how deficient, broken, diminished, and imperfect it may be, that in some way, in so far as it possesses a formal principle, does not, at the same time, have a soul, though it may not have the action of a subject that we call animal. And, in conclusion, it is agreed, with Pythagoras and the others who did not open their eyes in vain, that an infinite spirit fills and contains all, according to diverse manner and degree. Tenth, it is made to be understood, that since this spirit is permanent, along with matter which the Babylonians and the Persians called shadow—and since one and the other is indissoluble—that it is impossible, in any instance, for anything to be subject to corruption; nor can anything suffer death according to Substance; though, according to certain accidents, everything changes its aspect and is transmuted through one or another being, now in this, now in that composition; abandoning and retaking now this, now that being. Eleventh, how the Platonists, Aristotelians, and other sophists, have not recognized the substance of things; it is clearly shown how that which they call substance in natural things, aside from matter, is pure and simple accident. And, that from the cognition of the true form the true cognition of that which is life and death, is inferred. And, having completely destroyed the vain and puerile terror of this, a part of the happiness which accompanies our contemplation is known, in accordance with the foundations of our philosophy: since that philosophy pulls asunder the somber veil of insensate belief concerning Orcus and the avaricious Charon, with which we enrapture and poison the sweetest part of our life. Twelfth, form is distinguished, not according to its substantial nature, through which it is one, but according to the acts and exercises of the potential faculties and the specific degrees of beings which it produces. Thirteenth, the true definitive essence of the formal Principle is established; how the form constitutes a perfect species, differentiated in matter, according to its accidental dispositions—which depend on the material form, which consist of distinct grades and dispositions of the active and passive qualities. It is seen how the form is variable, how invariable, how it defines and determines matter, and how it is defined and determined by it. And last, it is shown, by an analogy, which is appropriate for the vulgar mind, in what manner this form, this soul, can be all in all, and in whatever part of the whole.

Argument of the Third Dialogue

The third dialogue—after having treated, in the first,* of the form, which has the nature of a Cause more than of a Principle—proceeds to the consideration of matter, which is judged to have more of the nature of Principle and element than of Cause. There, putting aside the preliminaries of the dialogue, it is shown first, that David of Dinant was not stupid, when he held that matter was an excellent and divine thing.

* The number (1), referring to the numbering of the dialogues, should read (2); The new reading is due to the fact that the first dialogue was added by Bruno, as an apology to "Le cena de le ceneri." See Chapter I, above.

Second, how through distinct ways of philosophizing, diverse concepts of matter can be apprehended, even though, in reality, there exists only one prime absolute matter; because it is realized in distinct grades, and is hidden under many such distinct species, different people can understand it differently, in accordance with the concepts that are appropriate to their doctrines; not otherwise (it follows) is it with number, which is understood by the arithmetician purely and simply; which is understood by the musician, harmoniously; by the Cabalist, symbolically; and in other diverse ways, by other madmen and by other wise men. Third, the meaning of the word, matter, is declared, by means of the difference and similarity that exist between the natural and the artificial subject. Fourth, it is proposed, in what manner the obstinate must be dispatched, and up until what point we are obliged to answer and discuss. Fifth, concerning the true essence of matter, it is inferred that no substantial form loses its being; and it is effectively demonstrated that the peripatetics and other common philosophers, in spite of the fact that they speak of substantial form, have never known any other substance than matter. Sixth, a formal, constant, Principle is established, in the same way that a constant material Principle is known; and that, with the diversity of dispositions that exist in matter, the formal Principle is transferred to the multiform configuration of distinct species and individuals; and it is shown, from whence it comes, that some, educated in the Peripatetic school, have not wanted to recognize any substance other than matter. Seventh, in what way it is necessary that reason should distinguish matter from form, and potency from act; and it is answered that it has already been shown above, in the second paragraph: that is, how the subject and principle of natural things, through different ways of philosophizing can be differently understood, without incurring falsity; but more usefully, according to natural and magical ways, and more variously, according to mathematical and rational ways; especially, if the latter conform in such a manner to the rule and exercise of reason that they do not effect in the end anything worthwhile, and do not obtain any fruit in practice, without which all contemplation must be considered to be vain. Eighth, two ways, in which matter used to be considered, are proposed: that is, how it is either potency, or subject. And starting with the first way: matter is distinguished, as active and passive potency; and in a certain manner, it is reduced to one. Ninth, it is inferred from the eighth proposition in what way the supreme and the divine is all that can be; and how the universe is all that can be; and other things are not all that they can be. Tenth, as a consequence of that which has been said in nine, it is shown, in a precise, profound, and evident manner: why there are in nature vices, monstrosities, corruption, and death. Eleventh, in what manner the universe is in no one of its parts, and in all the parts; and from this, proceeds to an excellent contemplation of divinity. Twelfth, why it happens that the intellect cannot comprehend this absolute act and this absolute potency. Thirteenth, the excellence of matter, which coincides with form, as potency with act, is declared. Last, as much from this, that potency coincides with act, and the universe is all that can be, as from other reasons, it is inferred that all is one.

Argument of the Fourth Dialogue

In the fourth dialogue, after having considered matter in the second* in so far as it is potency, matter is considered in so far as it is a subject. There, first, together with the amusements of Polyhymnius, the doctrine of matter is reported according to the common principles of some Platonists, as well as of all the peripatetics. Secondly, in discussing according to proper principles, it is shown, by many reasons, that the matter of corporeal and incorporeal things is one; of which reasons, the first is taken from the potency of the same genus; and the second, from the reason of certain proportional analogies between corporeal and incorporeal, absolute and contracted; and the third, from the order and ladder of nature, which rises to a first Principle, which reunites all in itself, and contains all; and the fourth, from the fact that there must be something indistinct, before matter is distinguished into corporeal and incorporeal, and that this indistinct being, is indicated by the supreme genus of the categories; and the fifth, from the fact, that just as there is a common concept for the sensible and the intelligible, so too there is for the subject of sensibility and intelligibility; and the sixth, from the fact that the being of matter is independent of the being of body, whence matter not with any less reason, can correspond to incorporeal and corporeal things; and the seventh, from the order of superior and inferior which is found in substances: for, wherever there is such gradation, there is presumed, and understood, a certain community (which is in accordance) with matter that is always represented by the genus, just as the form is represented by the specific difference; and the eighth, is taken from Principle, which is extraneous, but conceded by many; the ninth, from the multitude of species which is asserted in the intelligible world; the tenth, from the similitude and imitation of the three worlds, the metaphysical, the physical, and the logical; the eleventh, from the fact that, every number, diversity, order, beauty, and ornament,

* Bruno means the third; see above.

is about matter. Thirdly, four opposing reasons are briefly reported and are answered. Fourthly, it is demonstrated in what manner there exist diversities of nature, between this and that, between this and that matter; and how, in incorporeal things, matter coincides with act; and how all species of dimension exist in matter, and all qualities are comprehended in form. Fifthly, that no truly learned man has ever said that the forms are received from outside, by matter, but that they are sent out from within, as if they were taken from its bosom. Whence, it is not a "prope nihil," an almost nothing, a pure nude potency—if all the forms are, as it were, contained by it; and from itself, by virtue of the efficient cause (which can be even indistinct from it, in its being), produced and engendered; and they have not any lesser manner of actually existing in sensible and explicated being, though only in the way of their accidental subsistence, since all that which is seen and made manifest by the accidents founded on dimensions is pure accident and only the substance remains coincident with indivisible matter. Whence, it is clearly seen, that from explication, we cannot grasp any other thing than accidents; in such a way, that the substantial differences are hidden—according to Aristotle's statement (forced by the truth)—in such a manner, that, if we want to consider it well, we can well infer from this, that: the omniformed Substance is one, that truth is one, that being is one—a one, which according to innumerable circumstances and individual appearances, exhibits itself in such and such a diversity of subjects. Sixthly, how that which Aristotle and other similar ones understood by matter, "being in potency"—which certainly is nothing—is upheld without any reason; and according to these same ones, matter is, in fact, so permanent, that it never changes or varies its being, but every variety and mutation takes place about it; and that which "is," after having been potentially, (also for them) is always the composite. Seventhly, precisions are made concerning the appetite of matter, showing how vainly it is defined through that, without departing from the reasons (taken from the principles and suppositions) of those same ones, who proclaim it to be such as the daughter of privation, and similar to the insatiable greediness of the ardent female.

Argument of the Fifth Dialogue

In the fifth dialogue, which treats especially of the One, the foundation of the edifice, of the natural and divine knowledge, is completed. There, in the first place, the conception of the coincidence of matter and form, of potency and act, is established, in such a way, that "being," though logically divided into that which is, and that which can be, is really indivisible, indistinct, and one; and that it is at the same time, infinite, immobile, and indivisible, and without difference of part and whole, principle and principled. Second, that in that (One), the century is not different from the year, the year from the moment, the span from the furlong, the furlong from the mile; and in its nature, this and that other specific being is not other and other; and therefore there is no number in the universe, and therefore the universe is one. Third, that the point is not different from the body in the Infinite: because act and potency are not different things, and therefore, the point can run on at length, the line be extended in width, and the surface, in depth; the first is long, the second is wide, and the third is deep; and since everything is long, wide, and deep (consequently) they are one and the same, and the universe is all center and all circumference. Fourth, from the fact that Jove, (as they call him) is found to be more intimately in all, than the form of all can be imagined to be there (because he is the essence, through which all that is, has being; and since he is in all totally, everything has in itself the all, more intimately than its proper form), it is inferred that everything is in everything and, consequently, all is one. Fifth, answer is made to the doubt that demands to know: why all particular things change, and why the particular "matters," in order to receive one and another being, are forced to new and other forms; and it is shown how there is unity in multiplicity, and multiplicity in unity; and how being has many modes, and a multiple unity, and finally, that it is one in substance and truth. Sixth, inference is made concerning the derivation of that difference of particular things, and of that number of particular matters, and it is made explicit, that they are not being, but of being and about being. Seventh, it is marked that he who has found that one, that is to say, the reason of this unity, has found that key, without which it is impossible to enter into the true contemplation of nature. Eighth, with renewed contemplation, answer is made, that the One, the Infinite, the Being, and that which is in all, is throughout all, the selfsame, everywhere; just as the indivisible infinite multitude, not being "number," coincides with unity. Ninth, how, in the Infinite, there are no parts, parts being reserved for the unfolded universe; where, however, all that we see of diversity and difference are nothing but diverse and different aspects of the same Substance. Tenth, how, in the two extremes that are spoken of in the extremity of the ladder of nature, not two principles must be considered, but one; not two beings, but one; not two contrary and diverse principles, but one; concordant and identical. In it, height is depth; the abyss is the inaccessible light; obscurity is clarity; the great is the small; the confused is the distinct; strife is friendship; the divided is the indivisible; the atom is immense; and conversely. Eleventh, how, and in what way, certain geometrical denominations, like the point and the unity, are understood, in order to promote the contemplation of Being and the One; and how, they are not sufficient, of themselves, to express that. For which reason, Pythagoras, Parmenides, and Plato, must not be rashly interpreted, in accordance with the pedantic censure of Aristotle. Twelfth, from this: that Substance and being are distinct from quantity, from measure, and from number, it is inferred that it [Substance] is one and indivisible in all and in whatever thing that exists. Thirteenth, the signs and the proofs, through which is is shown, that contraries truly coincide, are reported; that they are derived from one principle, and that they are one in reality and Substance; all of which, is concluded to be true, physically, after having been shown to be true mathematically.

Behold then, most illustrious gentleman, from what point it is necessary to leave, in order to enter into that most special and most appropriate knowledge of things. Here, as in its proper seeds, is contained, and is implied, the multitude of conclusions concerning natural science. From here is derived the structure, the disposition, and the order of the speculative sciences. Without this introduction, it is vain to attempt, to begin, and to enter into that knowledge. Take then, willingly, this principle, this one, this fountain, this beginning, so that its seeds and its progeny may become stimulated to go out into the light of day; so that its rivulets, and many streams, may become diffused—multiplying their number successively, and disposing its members, each time further and further—so that, finally, with the ceasing of the night, with its somnolent veil and obscure mantel, the mighty Titan—parent of the divine muses, adorned by his family, and surrounded by his eternal court—brings forth the triumphal carriage from the red bosom of this gallant Aurora, (after banishing the nocturnal torches), adorning the world with a new day. Farewell.*

* Of the five poems which follow, three were written by Bruno in Latin, and two in Italian. The Latin poems are: "To the Principles of the Universe," "To My Own Spirit," and "To Time"; the Italian sonnets are: "Of Love," which is written in two parts. I have made no attempt other than to give a literal translation of these verses, for my purpose is to keep as close to the meaning as possible.

GIORDANO NOLAN

To the Principles of the Universe

He who, from his Laethaen origin, still abides in the flowing sea,
May ascend, O Titan, I implore, to the realm of the stars.
O erring stars, behold, I also begin the circular course, associated with you, if you opened the way.
Your movement may grant me that the double door of sleep be opened wide, when I rush up through emptiness.
What lies hidden long, unfavorably in thick veil, let me draw from dark night, into joyous light, to bring forth your fruit.
Why do you hesitate, feeble mind, though the time to which you lend your gift is unworthy?
However the shadows' swell covers the countries,
You, our Olympus, freely raise your head to the ether.

To My Own Spirit

Rootedly rests the mountain, deeply grown into one with the earth;
But its head rises to the stars.
You are kindred to both, my Spirit,
to Zeus as well as to Hades; and yet separated from both.
To Mind, a kindred mind calls you, from the height of things, that you should be the boundary between things above and below.
Do not sink, dumbly into Aneron's flow, low and heavy with dust.
No! Rather upward to heaven. There, search for your home.
For if a God touches you, you become a flaming Blaze.

To Time

Old One, who both limps and relaxes, who closes and opens up,
Is it more correct to call you good, or does one rather call you bad?
You give plenty and yet are stingy.
What you have given you steal; what you have borne, you will yourself destroy too;
What is taken from your bosom you swallow (down your throat).

If you begin everything, and destroy everything, in change,
May I not then call you good and bad at once?
Yet, where, in vain, you raise yourself to a cruel stroke,
There, do not stretch the menacing hand, armed with the scythe.
Where the last traces of the black chaos disappear
There, never show yourself good, never bad, O' Old One.

Of Love

——Love, through whom I perceive the truth so high,
Who opens the doors of diamonds and black ones for me,
My Deity enters through your eyes, and through seeing it,
Is born, lives, is nourished, and reigns eternally.
It shows show much the Heaven, the Earth, and the nether world contain, makes present the true images of absent things, recovers forces, and moving straight ahead, wounds,
And always injures the heart, and uncovers everything internal.
Therefore, vile populace, attend to the truth,
Listen to my words, that are not false.
Open! Open your eyes, insane and squinting, if you can.
You believe it to be childish, because you understand little;
Because at each instant you change, you think it is fugitive,
And through your being blind, you call it blind.

Cause, Principle, and One eternal
From whom being, life, and movement are suspended,
And which extends itself in length, breadth, and depth,
To whatever is in Heaven, on Earth, and Hell;
With sense, with reason, with mind, I discern,
That there is no act, measure, nor calculation, which can comprehend
That force, that vastness, and that number,
Which exceeds whatever is inferior, middle, and highest;
Blind error, avaricious time, adverse fortune,
Deaf envy, vile madness, jealous iniquity,
Crude heart, perverse spirit, insane audacity,
Will not be sufficient to obscure the air for me,
Will not place the veil before my eyes,
Will never bring it about that I shall not
Contemplate my beautiful Sun.

FIRST DIALOGUE

Interlocutors: *Heliotropio*, *Filoteo*, *Armesso*

Hel. Like criminals accustomed to the darkness, who, when freed from some dark tower, go out into the light, many of those trained in the common philosophy, and others too, will become frightened and awestricken, and being unable to endure the new sunlight of your clear conceptions, will become disturbed.

Fil. It will not be the fault of the light, but (the lack) of enlightenment; the more beautiful and excellent the sun may be, so much more hateful and unwelcome will it be to the eyes of the night witches.

Hel. The work which you have undertaken, Filoteo, is difficult, rare, and unique, since you wish to take us out of the darkness and lead us to the open, tranquil, and serene presence of the stars, which we see disseminated throughout the celestial mantel of the heaven with such beautiful variety. Although the helping hand of your pious zeal is succor to men alone, the reactions of the ungrateful against you will be none the less manifold; since various are the kinds of animals which the benign earth produces and nourishes in her extensive and maternal bosom; and since it is certain that the human species shows in its individuals the variety of all the other species; for there, the whole is present in each individual, more expressly than in the individuals of the other species.

For that reason, some, like the covered mole, will, just as soon as they feel the fresh air, look immediately for their native dark habitations again by furrowing in the earth; others, like the night owls, as soon as they see the dawning of the reddish ambassador of the sun in the illuminated east, will feel themselves attracted by their dark places of refuge because of the weakness of their eyes. All of the animals who are banished from the presence of the celestial lights and destined to the eternal cages, moats, and caverns of Pluto—when called back by the horrible horn of Alecto—will open their wings and direct themselves speedily to their lodgings.

But the animals born to see the sun, having arrived at the end of the hateful night, and being grateful for the goodness of the heaven, and making themselves ready to receive (in the center of the round crystal of their eyes) those so desired and so longed for rays, will revere the east with unusual adoration of heart, voice, and hand; and when, from the east's golden balcony, the beautiful Titan has released the fiery horses, and has broken the sleepy silence of the humid night, the men will talk, the docile, disarmed, and simple flocks of sheep will bleat, and the horned oxen will bellow under the vigilance of the rustic ox-herders; the horses of Silenus will bray in order to put fright again into the giants more stupid than themselves, thus helping the threatened gods; the teethed pigs, wallowing in their muddy beds, will deafen us with their bothersome grunts; the tigers, bears, lions, wolves, and the deceitful fox—sticking their heads out of their caverns and contemplating the open field of hunting from the deserted heights—will call forth from their savage breasts grunts, noises, roars, bellows, and cries.

In the air and on the leaves of the plants, (full of branches) the cocks, the eagles, the peacocks, the cranes, the turtledoves, the blackbirds, the sparrows, the nightingales, the crows, the magpies, the ravens, the cuckoos, and the cicalas, will not be tardy in repeating and re-echoing their stringent chirpings. And from the liquid and unstable element, water, the white swans, the painted ducks, the solicitous plungeons, the ducks which inhabit the lagoons, the hoarse geese, and the querulous frogs, will disturb our ears with their noises, in such a manner that the warm light of the sun, spread by the air of this fortunate hemisphere, will come to be accompanied, greeted, and perhaps molested by a variety of voices as great as the number and classes of the spirits who produce them out of the depth of their bosoms.

Fil. It is not only normal but also natural and necessary that each animal should emit its voice; and it is impossible that the beasts form regulated accents and articulated sounds like men, since the bodily constitutions are contrary, the tastes diverse, and the food distinct.

Arm. Please grant me the liberty of speaking. Not about the light, but about some circumstances, which are used, not so much to comfort the sense as to afflict the thought of him who sees and considers. For your own tranquillity and peace, (which with fraternal charity I wish you) I should not want these conversations of yours to be made into comedies, tragedies, laments, dialogues, or whatever else you might want to call them, similar to those that some while ago, (by having gone out into the open) have obliged you to remain hidden and secluded in your homes.

Fil. Speak freely.

Arm. I shall not speak as a holy prophet, nor as an absent-minded soothsayer, nor as an inspired Apocalyptic, nor as an angelic ass of Balaam; I shall not discuss as if inspired by Bacchus, nor as swelled with the wind through the prostituted muses of Parnassus, nor as a Sibyl impregnated by Apollo, nor as a prophetic Cassandra, nor as one invested from the tip of my toes to the top of my head with the enthusiasm of Apollo, nor as a seer illuminated by the oracle, nor the Delphic tripod, nor as an interrogated Oedipus faced with the difficulties of the sphinx, nor as Solomon before the enigmas of the queen of Sheba, nor as Calchas, the interpreter of the Olympic Senate, nor as a possessed Merlin, nor as one escaped from the cavern of Trophonius.

On the other hand, I shall speak in the common and vulgar language, like a man who has had some other thought than to go on lapping up the liquid of the great and small nape, to that point where there remains only dryness to the meninges (pia mater arachnoid, and dura mater). I shall speak, I say, like the man who has no other brain than his own, and whom even the heaviest drinking and eating gods in the celestial court (I refer to those who do not drink ambrosia, nor desire nectar, but to those who quench their thirst at the bottom of the barrel—and with the scattered wines—when they do not want to be taken care of by the nymphs and the lymphs; to those gods who are more domestic, cordial, and sociable with us), like Bacchus, or the drunken cavalier god, riding on an ass, Silenus, or Pan, or Vertumnus, or Faunus, or Priapus, who do not consider me worthy enough to give me one more little piece of straw, although such gods are used to heaping their favors even on horses.

Hel. This is much too long for a prologue.

Arm. Have patience; the conclusion will be brief. I want to say, in short, that I shall make you listen to words, which do not have to be deciphered—not having been distilled, nor passed through a retort, nor through a water glass, nor sublimated through the prescription of the quintessence. I shall speak such words as my nurse taught me—(my nurse) who was as greasy, as big bosomed, as big bellied, as big hipped, as big rumped, as only that Londoner was, whom I saw in Westminster, who (for the heating of her stomach) had such big breasts that they seemed to give the appearance of the half boots of the giant San Sparagorio, and from which (it seemed that) leather stitching could make two large tenarese bagpipes.

Hel. It seems to me that this should be a sufficiently long prologue.

Arm. Well, then, in order to come to the end of all this, I should like to find out from you (purposely leaving to one side now the voices and the tongues, for the sake of regarding the light and the splendor, which your philosophy may bring us) with what voices do you want us to greet, in particular, that splendor of doctrine which issues from the book, "La cena de le ceneri"? What animals are those that have recited of the "La cena de le ceneri"? I ask—are they aquatic, aerial, earthly, or lunar? And leaving to one side the reasoning of Smith, Prudenzio, and Frulla, I want to know if those are in error, who affirm that there, you play the part of the rabid dog, without any prejudice to playing also, the part of now, the monkey, now, the wolf, now, the magpie, now, the parrot, now, this animal, now, that animal—mixing all sorts of discourses, grave and serious, moral and natural, noble and ignoble, philosophical and comical?

Fil. Do not marvel, brother, because that was only a supper, where brains are governed by the emotions prompted by the effect of the flavors and smells of the food and drink. In conformity, then, with the way, in fact, that the supper is corporeal and material, it will consequently be verbal and spiritual. In this way, this dialogued supper has its variously distinct parts, just as that other (corporeal) one is accustomed to have; and not otherwise, this also has its conditions, circumstances, and recourses, just as the other one might have its own.

Arm. Please enlighten me.

Fil. At the corporeal supper, as is usual and desirable, there are to be found salads and nutritious foods, fruits and the usual things, kitchen specialties and spices, things for the healthy and for the sick, cold dishes and warm ones, raw greens and boiled meats and vegetables, sea food and earth food, cultivated foods and venison, roast meats and boiled meats, ripe foods and green ones, some nutritious foods, and some for the pure taste of the palate, substantial things and light ones, piquancies and insipids, sour foods and sweet ones, bitters and delicacies.

In the same way, and as a certain consequence, at the other supper, there have appeared oppositions and controversies, adapted to the diverse stomachs and tastes which may want to present themselves at our ideal symposium: to the end that there may not be anyone who can lament the fact that he has arrived in vain and to the end that that one to whom one serving is not pleasant, may be served with another.

Arm. That is certain; but what would you say if besides these there appeared at your banquet and your supper things that are not good, neither as salad nor as nutritious foods; nor as fruits, nor as common things at dinner; neither as cold foods, nor as warm ones; neither as raw greens, nor as boiled foods; things that are not good, neither as appetizers, nor as satisfactions for hunger; things that are not good, neither for the healthy, nor for the sick; things, finally, that would be better liked, if they had never left the hands of a cook or an apothecary.

Fil. You will see that our supper is not different, in this, from whatever other supper that can be given; since, just as in another supper, it may well be that, at the best moment of eating, either you will encounter a morsel that is so hot at the point of going down that you are obliged to expel it immediately, or to send it down slowly—while crying and tearing—through the palate, until you can give it that cursed thrust down through the throat; or it may well be that some tooth begins to pain you sharply; or you may have your tongue intercepted, so that you come to bite it along with the bread; or you may have a small stone (come to) break and steal itself in between your teeth, so that you have to return the mouthful; or, there may well be some hair from the cook's head stuck on your palate, to the degree that you almost vomit; or, there may well be a fish bone stuck in your throat, and it makes you cough heavily; or, it may well be, that some bone has penetrated the throat to the point where it places you in danger of choking; so also, at our supper—to our common disgrace—it may well be: there have been found things that correspond and are proportional to those of the other supper. All of this happens on account of the sin of the first man, Adam; on whose account perverse human nature is condemned to have the disgusting joined to the agreeable.

Arm. You have spoken piously and saintly. Well now, what do you answer to what they say, namely, that you are a fanatical cynic?

Fil. I concede that easily; if it is not completely so, it is partly true.

Arm. But you know that it is less dishonorable for a man to receive outrages than to commit them?

Fil. But it is sufficient that those that I commit may be called vengeances, and those committed by the others, offenses.

Arm. Even the gods are exposed to injury, are liable to defamation, and are subject to vituperation; but to vituperate, to defame, and to injure, is a property of cowardly, ignoble, vulgar, and malicious men.

Fil. That is true, but we are not injurious; we renounce the injuries which are done, not so much to ourselves, but to the despised philosophy; in order that to the unpleasantries already received others will not be added.

Arm. Do you want, then, to appear to us like a dog who bites, so that no one dares to molest you?

Fil. So it is; for I like calm, and I detest displeasure.

Arm. Yes, but it is the accepted opinion that you proceed with too much rigor.

Fil. In order that they do not do the same again, and the rest learn not to dispute with me and with others; treating these conclusions with similar half measures.

Arm. The offense was private and the vengeance is public.

Fil. But it is not for that reason unjust; because many errors are committed in private, that are justly punished in public.

Arm. But by proceeding in this manner, you will hurt your reputation; and you make yourself more blameworthy than those, because it will be said publicly that you are impatient, fantastic, arbitrary, and frivolous.

Fil. I do not care, provided that they, or any others, do not bother me; for that reason, I show the cynical stick; in order that they allow me to remain at peace; and if they do not want to make overtures to me, they should not exhibit their ill feeling toward me.

Arm. And does it seem to you to befit a philosopher to want to avenge himself?

Fil. If those who molest me were a "Xanthippe," I would be a Socrates.

Arm. Do you not perchance know that long sufferance and patience are becoming to all, and that because of them men become (like) heroes and great gods, who, according to some, take revenge after great delay; and according to the opinion of others, never revenge themselves; nor are they provoked to anger?

Fil. You deceive yourself, if you believe that I have ever thought of avenging myself.

Arm. What else then?

Fil. I have intended to correct them; in this exercise, we too are like the gods. You know that poor Vulcan has been obliged by Jove to work even on holidays, and that that cursed anvil never tires of bearing the blows of so many fierce hammers—in such a manner that as soon as one of the hammers is raised, the other is already lowered—so that the just flashes of lightning, with which the delinquent and the criminal are punished, should never be lacking.

Arm. There is a difference between you and the smith of Jove, and the husband of the goddess of Cyprus.

Fil. It is sufficient for me not to be different from those gods, even in the patience and the long suffering, which I have exercised in my behavior—that is, is not completely loosening the brake on my anger nor in not more strongly inciting my ire.

Arm. It does not fit everybody to correct others, especially the multitude.

Fil. Say still more; especially when the multitude does not yet touch him.

Arm. It is said that one must not be a reformer in a foreign country.

Fil. And I assert two things: first, that one must not kill a foreign doctor, because he intends to perform those cures which those of this same country do not perform; and second, I declare that, to a true philosopher, every land is his fatherland.

Arm. And what if they do not accept you, neither as a philosopher, nor as a doctor, nor as a compatriot?

Fil. I will be so nevertheless.

Arm. Who assures you of that?

Fil. The gods, who have placed me here; I, who find myself here; and those who have eyes and see me here.

Arm. Your witnesess are few and not well known.

Fil. Real doctors are few, and little known; almost all are really ill. And I repeat that they have no right, the ones to cause, and the others to permit, that such treatment be given to those who offer such honorable merchandise; whether they are strangers or not.

Arm. There are very few who recognize that merchandise.

Fil. The gems are no less precious on that account; and we must with all our power defend them and make them defend, liberate them, and free them, from being trampled under the feet of the hogs. And so may the gods be favorable to me, Armesso, as it is certain that I never committed such vengeances, neither through sordid love of myself, nor through a base care for some particular lesson; but I did so only through my love of "my so beloved mother philosophy" and through zeal for her offended majesty, which, on account of her false friends and sons (because there is no vile pedant, idle phrasemaker, stupid faun, or ignorant horse, who does not wish to be considered one of the family, either by exhibiting himself loaded down with books, or by making himself grow a beard, or by other apparent means of masquerade) is reduced to such a state that among the common people a philosopher is tantamount to an imposter, an idler, a common pedant, a faker, a charlatan—good because he serves as amusement in the house and as a scarecrow in the country.

Hel. To tell the truth, the family of philosophers is considered by the majority of people to be more contemptible than the family of chaplains, because the latter, who have been adopted by all kinds of bad people, have not made the priesthood as contemptible as the former, (named after all kinds of beasts) who have put philosophy in contempt.

Fil. Let us praise, then, in its manner, antiquity; when philosophers were such, that from among them there were chosen those who were promoted to legislators, counsellors, and kings; when such were the counsellors and kings, that they were elevated from this state to that of the priests. In these times, the greater part of the priests are of such a kind that they are despised, and the divine laws are also despised on their account; such are almost all those philosophers whom we see; they are despised and the sciences are also despised on their account. We find, moreover, a number of crooks among these, who, like nettles, are accustomed to oppress (with their contrasting dreams) that rare virtue and truth, which manifest themselves only to a few.

Arm. I do not find a philosopher, who is so irritated, by seeing philosophy vilified, nor, O Heliotropio, do I encounter anyone so fond of his science as this Teofilo; what would happen if all the rest of the philosophers were in the same state—that is, so impatient?

Hel. Those other philosophers have not found as much as he, nor do they have so much to care about and so much to defend. The others can easily vilify that philosophy which is of no value; or another that is worth very little; or that which they do not know; but "that" one who has found the truth, which is a hidden treasure—when he becomes enamored with the beauty of that divine face—becomes just as jealous as anyone else (can become) of his gold, of his rubies, of his diamonds, or of a carrion of feminine beauty, because that divine face is not to be defrauded, depreciated, nor contaminated.

Arm. But let us return to our thesis; let us come to the "why." They say of you, Teofilo, that in your "Cena," you criticize and do injustice to an entire city, to a whole province, and to an entire kingdom.

Fil. I have never thought of that, nor have I ever done that, nor have I ever had the intention of doing that; and if I might have thought of doing that, or sheltered the intention of doing that, or if I had done that, I would condemn myself for the worst; and I would be prepared for a thousand retractions, and a thousand recantations, a thousand denials—not only if I had injured a noble and ancient kingdom, like this one, but any other one, however barbarous; if I had done injury not only to any city, however uncivilized, but to any race, however savage; or to any family, however inhospitable; because there cannot be a kingdom, city, or race, nor an entire house, that can be or must be supposed to be of an identical condition; and where, there may not be opposite and contrary customs; in such a way, that that which is agreeable to one may not be disagreeable to another.

Arm. Certainly; in so far as I am concerned, who has read and re-read, and re-examined everything very well, though about particulars somehow I find you somewhat too diffuse; in general, I find you proceeding rigorously, reasonably, and thoughtfully; but rumor has spread it in the manner in which I have been reporting it to you.

Hel. That rumor, and many others, have been spread through the meanness of some of those who feel that they have been hurt; and desirous of vengeance, and feeling themselves without sufficient reasons, doctrine, ingenuity, and power—apart from inventing all other falsities that they can—in which only others like them can believe, look for company by trying to suggest that the particular punishment which they receive may appear to be a common injury.

Arm. I believe, on the other hand, that there are persons who do not lack judgment and council, and who still consider the injury to be universal, because you manifestly place such customs among the people of this nation.

Fil. But what are those mentioned customs, that similar, or worse and sharper ones—in kind, species, and number—may not be found in the most excellent places, territories, and provinces of the world? Would you call me, perchance, injurious and ungrateful to my country if I said that similar and more criminal customs are to be found in Italy, in Naples, and in Nola? Would I, perhaps, belittle that region, blessed by heaven, that is placed at the head and at the right hand of the globe, that governor and ruler of all the other nations, which has always been esteemed by us and by others to be master, nurse, and mother of all the virtues, disciplines, humanities, modesties, and courtesies—if one might come to exaggerate even that which our own poets have sung concerning it, our poets, who make it no less than the master of all the vices, deceits, avarices, and cruelties?

Hel. All that is certain, in keeping with the principles of your philosophy, according to which you want contraries to coincide in the principles, and in the related subjects. The same minds, which are endowed with great aptitude for highly virtuous and generous undertakings, end up in extreme vices, if it happens that they become perverse—to say nothing of the fact that there are apt to be found there rarer and more select minds—where the majority are very ignorant and stupid, and where, in general, they are less cultured and less courteous—and, in particular it is customary to find extreme urbanity and courtesy, in such a way, it seems, that the many nations are given an equal proportion of perfection and imperfection, albeit in different ways.

Fil. You speak the truth.

Arm. And yet, Teofilo, it pains me, and with me many others, that in our lovely country you have encountered such subjects who have invited you to an ash dinner in order to have the occasion to complain to you, rather than those many others who have made it evident to you to what degree our country—though it has been declared by your compatriots to be "penitus toto divisus ab orbe"—is inclined toward all the studies of good letters, arms, chivalry, humanities, and courtesies;

in all of which, and as far as our powers permit it, we exert ourselves, so as not to be less than our ancestors, and so as not to be conquered by other nations, especially by those that are considered to have nobility science, arms, and any other civilities by nature.

Fil. On my faith, Armesso, with regard to what you have said, I ought not and should not know how to contradict you, neither with words, nor with arguments, nor with my conscience—since you defend your case with every argumentative skill and with every modesty. And, therefore, because of you, that is, because you have not approached us with barbarous pride, I begin to repent and to regret that I ever received the impetus from those others to attack you and others who are of an honorable and humane character. And, therefore, it would please me very much if I had never written those dialogues; and if you wish, I shall exert myself in order that henceforth they shall not reach the public eye.

Arm. My grief, together with that of other noble persons, derives so little from the circulation of those dialogues that I would willingly strive to have them translated into our language so that they would serve as a lesson to the poorly and badly educated among us, who perchance when they see the revolt that they produce, and the manner in which their crude reunions are described, and the bad impression that they make, might bring it about that, though they might not change their ways and follow the example of the greatest and the best, at least they might feel moved to imitate them if only for the shame of being reckoned among the others. They would thus learn that honor and valor do not consist in knowing and in being able to be troublesome in that manner, but precisely the other way around.

Hel. You show yourself to be very understanding and very clever in defending the cause of your country, and you are not ungrateful for the good offices of others, as is usually the fault of those who are weak in arguments and council. But it seems to me that Filoteo is not so clever in preserving his reputation and defending his person; for, in the same manner that nobility and rusticity are different, opposite results are to be expected and feared from them; for example: from a villain from Scythia who will become clever and who will be celebrated for his success if by leaving the banks of the Danube and going to tempt the authority and majesty of the Roman Senate with just complaint and bold censure, he will bring it about that the Senate will rise on this occasion of his censure and blame to take some action of extreme prudence and magnanimity and honor him with a statue; on the contrary, a Roman gentleman and senator—to judge by the bad result that he would obtain—would not in the least become wise, if abandoning the pleasant shores of his Tiber, he goes to test the villainous Scythians, even though it is with just complaint and reasonable censure; they will find an occasion, in this, to build real Babylonian towers of arguments of greatest villainy, infamy, and roughness; and they will stone him, releasing the brake to popular fury, in order that the other nations will better come to know the difference that there is in dealing and living with human beings and with those who are merely made to the image and likeness of men.

Arm. It must never come about, Teofilo, that I ought to, or could judge it proper that I—or any other more ingenious than I—should want to assume the defense and protection of those who are the object of your satire, because it might concern the people and the persons of this country to whose defense we are led by natural law itself; for I will never admit and I will always remain the enemy of anyone who affirms that they are parts and members of our country, which is made up of no other than persons, as noble, cultured, honest, disciplined, discreet, human, and as reasonable, as those of any other nation. Hence, although they are within the country, they are found there to be nothing else but the dirt, dross, excrement, and carrion; they could only be called part of the kingdom or the city in the same way that the hold is called a part of a ship. For these reasons, we should not resent them; for resenting them, we would be a thing to be condemned. I do not exclude from them the majority of doctors and clerics, although the doctorate makes gentlemen of some of them; and for the rest, of course, it serves the favor of petulancy and presumption, which gives them the reputation of cleric or literateur and serves to exhibit, with cynical emphasis, the rude sufficiency which they would not dare to manifest before; for this reason, you must not be surprised to see that many who have such a doctorate and an order of the priesthood know more about cattle, herds, and stables, than the grooms, ox-herders, and goat-herders themselves. Because of this, I would have preferred not to hear you speak so harshly of our University, that is, in terms of not pardoning it, even in general; and without taking into consideration what it has been and what it will or may be in the future; and, in part, what it is at present.

Fil. Do not feel disturbed on that account because, even though that university has been presented in detail, on this occasion, still, its errors are not such that the other universities do not likewise commit; that is, those universities which are considered the greatest and from which are graduated in most cases—with the title of doctors—bejeweled horses and asses with diadems. I do not deny, however, the wonderful organization which this university has established since its origin; the beautiful order of its studies, the gravity of its ceremonies, the arrangement of its exercises, the decorum of its habits, and many other details, which are necessary, and belong to an Academy; hence, without any doubt, everyone must affirm that it is the first in all Europe; and, consequently, the first in all the world. And I do not deny, that on account of the subtlety of the minds, and the keenness of the geniuses, which one and another part of Britain naturally produces, is similar to, and can be equal to, all those that are truly excellent. Nor is the memory of that fact forgotten that, before the speculative sciences were found in the other parts of Europe, they flourished here; and by means of these principles of metaphysics, though barbarous in language and, as it were, by professional clerics, the splendor of a rare and noble part of philosophy (which in our day is almost extinct) was propagated through all the academies of the non barbarous regions.

But what has troubled me, and what has occasioned for me, at times, laughter and disgust, is the fact that (whereas) I do not find anywhere, people more Roman and more Attic in language than I find here; otherwise (I speak in general) they pride themselves in being entirely distinct and even contrary to those who preceded them—that is, those who, caring little for eloquence and grammatical rigor, attended exclusively to speculations, which are now called sophisms by them. Yet I appreciate more the metaphysics of those, wherein they have surpassed their leader Aristotle—although it was a foul metaphysics and spoiled by vain conclusions and theories, which are neither philosophical nor theological, but proper only for the idle and badly employed minds—than that which those of the present age can contribute, with all their Ciceronian eloquence and art of declamation.

Arm. Those things are not to be depreciated.

Fil. Certainly; but if one has to make a choice between the two things, I esteem the culture of the mind, as sordid as it may be, more than all the most eloquent words and languages.

Hel. That reminds me of Brother Ventura; commenting on a passage from the Holy Gospel which says: "Render unto Caesar that which is Caesar's," he gathered together all the names of all the coins used in Roman times, together with their markings and weights—I do not know from what devilish annals or memorandum book he took them, and there were more than 120 of them—but he did all of this to show how studious and retentive he was; and having finished his sermon, a good man approached him and said: "My dear reverend father, please lend me a coin"; to which the good friar answered that he belonged to the mendicant order.

Arm. And what do you propose to show by all this?

Hel. I mean to say, that those who are well versed in phrases and names, and do not preoccupy themselves with things themselves, ride the same mule as that reverend father of mules.

Arm. I believe that besides the study of eloquence, in which they excel all of their predecessors, they are not inferior to the other moderns; neither are they destitute in philosophy and the remaining speculative studies; and without training in them, they cannot be promoted to an academic degree; therefore, the statutes of the university, which they are obliged to respect under oath, establish that no one is to be promoted to the degree of master and doctor of philosophy and theology unless he will have drunk from the fountain of Aristotle.

Hel. I shall, therefore, tell you what they have done, in order not to be perjurers. Of the three fountains in the university, they have given one the name of Aristotelian fountain; another, the name of Pythagorean fountain; and the third, the name of Platonic fountain. And just as the water which is used to make beer and ale is taken from these three fountains (the same water in which the oxen and the horses also water themselves), so it comes about that there is no one who, having remained in those colleges and in those study halls for three or four days, does not become impregnated, not only with the Aristotelian fountain, but also with the Pythagorean and with the Platonic.

Arm. Aye, what you say is unfortunately true; and so it happens, Teofilo, that doctors are as cheap as herrings; for just as they are bred, found, and fished with very little effort, so they are also bought at a low price. Such being, then, the multitude of doctors among us, in this day and age—though saving the prestige of some who are celebrated for their eloquence, their doctrine, and for their exceptional civil courtesy, for example, a Tobias Matthew, a Culpepper, and others whom I cannot mention—it has come about that having the title of doctor, far from being entitled to a new grade of nobility, is rather suspected to be of a contrary nature and condition, unless he is especially known. And so it happens that those who are noble by birth or through any other circumstance (even when that which has come to be the principal part of nobility is added to them, through knowledge) are ashamed to graduate and to be called doctors; hence, they satisfy themselves with "being learned." And you can find a greater number of these in the courts than pedants in the university.

Fil. Do not, therefore, complain, Armesso; because in all the places where there are doctors and priests, one and another seed of them is found. Whence those who are true doctors and true clerics—even though they have been elevated from a humble condition—cannot help but become ennobled and civilized, since science is an excellent way to make the human soul heroic. But the others show themselves to be the more rustic, in a very evident way, the more they desire to "thunder" either with the father of the gods or with the giant Salmoneus, when they "walk around" like a purpled faun or satyr (with that horrible, imperial masquerade) after having determined (from a professional chair) to what declension, "hic, haec, et hoc nihil" belong.

Arm. Let us now put aside these things. What book is that which you are carrying in your hand?

Fil. They are certain dialogues.

Arm. The "Cena"?

Fil. No.

Arm. What then?

Fil. They are others in which we treat the Cause, Principle, and One in accordance with our doctrine.

Arm. Who are the interlocutors? Perhaps we shall have another devil, like Frulla, or Prudenzio, who will engulf us in some new trouble.

Fil. Do not doubt that with the possible exception of one the rest are peaceable and very honest people.

Arm. Could it be that according to what you say we have yet to remove some thorny peels from these dialogues?

Fil. Do not doubt, for you will be scratched where it itches, rather than stung where it hurts you.

Arm. Now?

Fil. Here, you will meet first that learned, honest, affectionate, cultured, and very faithful friend, Alexander Dixon, whom the Nolan loves as his own eyes, and who has made it possible for this thesis to be planned. He is introduced as the one who proposes the subject matter to Teofilo. Secondly, there is Theophilus, that is, I, who serves to distinguish, to define, and to demonstrate the material in question, as it may fit the occasion. The third is Gervasius who is not a philosopher by profession, but who wants to be present at our conversations (and serves as pastime); he is a person who, as it were, neither stinks nor smells; he makes jokes of the things that Polyhymnius says, and he is the one who, from time to time, makes Polyhymnius manifest his frenzy. This sacrilegious pedant is the fourth interlocutor; he is one of the most rigid censors of philosophers; he is the one through whom Momus asserts himself; he is very much attached to his group of students, and for this reason, he is called "Socratic" in his love; he is the perpetual enemy of the female sex, for which reason, in order to be impersonal, he considers himself like Musaeus, Tityrus, and Amphion. He is one of those who, when they have made a "beautiful construction," or produced an elegant "little letter," or stolen a beautiful phrase from the Ciceronian kitchen, acts as if Demosthenes has been revived, as if renewed life has been given to Tullius, and as if Salustius has been vivified; here is an Argus, who sees each letter, translates each syllable, each word; here Rhadamathus, "calls the shadows of the silent (dead)"; here Minos, king of Crete, "moves the urn." They examine the orations; they discuss the sentences saying: these sound like those of a poet; these, like those of a comic author; this is serious; that is light; this is sublime; that is the humble kind of speaking; this oration is rough: it would be smoother, if it were formed thusly; this is a "beginner," little versed in the "ancients"; "he does not sound like Cicero, he lacks knowledge of Latin"; this word is not Tuscan, nor is it used by Boccaccio, Petrarch, and other approved writers; one does not write "homo," but "omo"; not "honore," but "onore"; not Polihimnio, but Poliinnio; and with this, he triumphs, he is satisfied with himself; what he does, pleases him more than anything else. He is a Jove, who, from his raised observatory, contemplates and considers the life of the other men, who are subject to such errors, calamities, miseries, and useless anxieties. Only he is happy; only he lives a heavenly life, when he contemplates his divinity in the mirror of a "Spicilegium," a "Dictionary," a "Calepinus," a "Lexicon," a "Cornucopia," a "Nizzolius." In possession of this sufficiency, he alone is everything, while everyone else is one only. If he happens to laugh, he calls himself Democritus; if he happens to be pained, he calls himself Heraclitus; if he discusses, he calls himself Chrysippus; if he discourses, he calls himself Aristotle; if he imagines, he calls himself Plato; if he ruminates on a little sermon, he is Demosthenes; if he analyzes Vergil, he is himself, Maro; he corrects Achilles, approves Aeneas, reprehends Hector, exclaims against Pyrrhus, condoles himself with Priam, argues against Turnus, excuses Dido, and commends Achates; and finally, while he renders word for word, and accumulates his savage synonyms, he believes that nothing divine is foreign to him. And he comes down from his cathedral, as haughty as if he had just ordered the heavens and directed the Senate, or mastered armies and reformed the worlds; assured that, were it not for the misery of the times, he would achieve, in fact, what he does in thought. *O tempora O mores!* How few are those who understand the essence of participles, of adverbs, and conjunctions! How long a time has passed since one has not found the reason for, and the true cause on account of which the adjective must agree with the noun, the relative with the antecedent; and according to what rule it is placed now at the beginning, and now at the end of a sentence; and in what measure and order, those interjections: "dolentis, gaudentis, heu, oh, ahi, ah, hem, ohe, hui," (and others) are to be interjected; without which the whole discourse is completely insipid?

Hel. Say what you will, and voice whatever opinion you please on that score; but I assert that in order to be happy in this life, it is better to feel yourself a Croesus, while being poor, than to consider yourself poor, while being a Croesus. Is it not, perchance, more convenient for happiness to have a "pumpkin" who appears beautiful and is pleasant to you than to have a Leda or an Helena, who bothers you no end, and brings you to weariness? What does it matter to them, as such, to be ignorant and ignobly occupied, if they are so much the more happy, the more they gratify only themselves? Hence, the fresh herb is as good for the ass, and the oats are as good for the horse, as the white bread and partridge are for you; hence too, in the same way, the pig is as contented with his acorns and broth, as a Jove with his nectar and ambrosia. Do you want, perchance, to separate these from their sweet folly, in such a way, that on account of that cure, they must then break their heads? I leave aside the question: Who knows whether this or that is folly? A Pyrrhonist has said: Who knows whether our state is not a state of death, and that which we call death is nor really life? In the same way, who knows whether all felicity and true beatitude does not consist in the dutiful union and the combination of the members of an oration?

Arm. So is the world constituted: we play Democritus on the pedants and the grammarians; the solicitous courtesans play Democritus on us; the monks and the priests, who are little occupied with thought, play Democritus on all; and reciprocally, the pedants ridicule us; we, the courtesans; and all, the monks; and in conclusion, while each one is a fool to the other, we all become different in species, and concordant in "kind, number, and case."

Fil. The kinds and modes of censure are thus various, and various too are its degrees; yet the roughest, hardest, and most horrible and most terrible, are those of our church teachers. Yet we must kneel and bow our heads before them; we must turn our eyes and raise our hands, sighing, crying, exclaiming, and asking mercy. I turn, therefore, to you, who carry in your hands the herald's staff of Mercury; so that you may decide the controversies and judge the questions which are stirred up between the mortals and the gods; to you, O Menippi, who, seated on the globe of the moon, look down at us with torturous glances, with repugnance, and with abhorrence for our acts; to you, shield bearers of Pallas, standard bearers of Minerva, stewards of Mercury, merchants of Jove, brothers of Apollo, highwaymen of Epimetheus, cupbearers of Bacchus, caretakers of the Bacchants (asses of Evantis), scourgers of the Hedonides, comforters of the Thyriads, seductors of Manades, bribers of the Bassarides, horsemen of the Mimallonides, consorts of the nymph Egeria, moderators of enthusiasm, demogogues of the errant people, discus bearers of Demorgorgon, gods of the fluctuating disciplines, treasurers of Pantamorphus, and emissary rams of the high priest Aaron; to you, we recommend our prose; we submit our muses; our premises; our digressions; our parentheses; our applications; our clauses; our periods; our constructions; our adjectives; and our epithets. Refer our barbarisms to good council, attack our solecisms, castrate our Silenuses, put pants on our Noahs, make eunuchs of our macrologies, patch up our ellipses, curb our tautologies, moderate our sharpnesses, pardon our indecencies, excuse our meanderings, and pardon our obscenities—you smooth aquatics, who, with your beautiful elegancies, steal our souls, fasten youselves to our hearts, fascinate our minds, and place in brothel our meritricious souls. I again exorcise you all, in general, and you Polyhymnius, in particular. I ask you, severe, supercilious, and savage master, Polyhymnius, to give up that furious rage and criminal hate of the noble female sex; do not disturb that which is the greatest beauty in all the world, and that which the heaven contemplates with her innumerable eyes. Return to yourself, recover the intelligence, with which you can see that this, your rancor, is nothing but expressed folly and fanatical zeal. Who is more insensate and stupid than the one who does not see the light? What folly can be more despicable than that which is, on account of sex, an enemy of nature itself; just as that barbarous king of Sarza, who, having learned it from you, said:

> Nature will not make anything perfect
> Since, concerning woman, its name is correct.

Contemplate the truth; raise your eyes to the tree of the knowledge of good and evil; see the contrariety and the diversity that exists between one and the other. Consider what men are, and what women are; for example, here you see (as a substratum) the body, your friend, being male; there, the soul, which is your enemy, being female; here you have chaos, male; there, order, female; here sleep, there vigilance; here lethargy, there memory; here hate, there love; here fear, there assurance; here rigor, there gentleness; here scandal, there peace; here tumult, there tranquillity; here error, there truth; here defect, there perfection; here hell, there happiness; here the pedant Polyhymnius, there the muse Polyhymnia; and to summarize: all the vices, defects, and crimes, are masculine; and all the virtues, excellencies, and goods are feminine; hence, prudence, justice, fortitude, continence, beauty, majesty, dignity, and divinity, are called feminine; thus they are imagined, thus they are described, thus they are represented, and thus they are.

And now, leaving these theoretical, conceptual, and grammatical reasons, which concern your subject, and coming to the real, natural, and practical proofs: should not this example alone, have been sufficient to bridle your tongue, stop your wagging mouth, and to confuse you, together with all those who think like you; that is, if one should ask you to find a man who was better, or at least similar to the divine Elizabeth, who reigns in England; who is endowed by heaven, and exalted, favored, protected, and sustained, in such a manner that the words and forces of others will endeavor in vain to dethrone her. I say to that lady, than whom there is none in all the kingdom more worthy; than whom there is none more heroic among all the nobility; than whom there is none more learned among the gowned; than whom there is none more prudent among the counsellors? Compared with her—as much for her bodily beauty as for her knowledge of the learned and common languages, as much for her knowledge of science and art, as much for her prudence in government, as much for her success in the exercise of the great and lasting authority, as much for the other civil and natural virtues—the Sophonisbas, the Faustinas, the Semirasmises, the Didos, the Cleopatras, and all the rest of those of whom Italy, Greece, Egypt, and the other territories of Europe and Asia can pride themselves, are very vile indeed. My evidence is: the effect and the brilliant success which this century reflects; and not without wonder when over the soil of Europe run the angry Tiber, the threatening Po, the violent Rhone, the bloody Seine, the turbulent Garonne, the tumultuous Ebro, the furious Tago, the agitated Meuse, the restless Danube—she, in the space of tewnty-five years and more, has, through the splendor of her eyes, calmed the great Ocean, which, with its constant ebbing, happily and tranquilly receives the beloved Thames in its great bosom; and which, free from fear and care, runs confidently and happily, undulating between the shores covered with herbs. Returning now—to start with the heading—what . . . .

Arm. Enough! Enough! Filoteo. Do not exert yourself to add water to our ocean and light to our sun. Stop discoursing so much in the abstract, not to say worse, in your polemic against the absent Polyhymniuses; show us something of those present dialogues, so that we do not spend the entire day, and these hours, idly.

Fil. Take them and read.

SECOND DIALOGUE

Interlocutors: *Dixon*, *Theophilus*, *Gervasius*, *Polyhymnius**

Dix. Have the kindness, Master Polyhymnius, and you too, Gervasius, not to interrupt our discourse further.

Pol. So be it.

Gerv. If he who is the master speaks, then obviously I cannot keep silent.

Dix. Then you say, Theophilus, that everything which is not a first principle and a first cause, has a principle and a cause?

Theo. Without any doubt and without any controversy.

Dix. Do you believe, then, that whoever knows the things so caused and originated knows the cause and the principle?

Theo. Not easily the proximate cause and the proximate principle, and with the greatest difficulty, the first cause and the first principle, even in its traces.

Dix. Then, how do you think that those things, which have a first and a proximate cause and principle, can truly be known, if their efficient cause (which is one of the things which contribute to the true knowledge of things) is hidden?

Theo. I submit that it is easy to set forth the demonstrative theory, but the demonstration itself is difficult. It is very easy to set forth the causes, circumstances, and methods of the sciences, but afterwards our method makers and analytical thinkers use but poorly their *Organa*, the principles of their methods, and their art of arts.

Gerv. Like those who know how to make excellent swords, but do not know how to use them.

Pol. *Ferme!*

Gerv. May your eyes be closed, so that you may never open them.

Theo. I say, therefore, that one should not demand that the natural philosopher make plain all the causes and principles, but only the physical, and of these only the principal and proper ones. Although, then, because they depend upon the first principle and cause, they can be said to possess that cause and that principle, this is, nevertheless, not such a necessary relation, that from the knowledge of one, the knowledge

* These same interlocutors appear in the dialogues that follow.

of the other could be inferred; and, therefore, one should not demand that in the same scientific discipline both should be set forth.

Dix. How is that?

Theo. Because from the knowledge of all dependent things, we cannot infer other knowledge of first principle and cause than by the less efficacious method of traces; for all things are derived from his [God's] will or goodness, which is the principle of his operation, from which proceeds the universal effect. The same situation arises in the consideration of artificial things, inasmuch as he who sees the statue does not see the sculptor; he who sees the portrait of Helen does not see Apelles, but he sees only the result of the operation, which comes from the excellence of the genius of Apelles; the work is entirely an effect of the accidents and circumstances of the substance of that man, who, as regards his absolute essence, is not at all known.

Dix. So that to know the universe is like knowing nothing of the essence and substance of the first principle, because it is like knowing the accidents of accidents.

Theo. Correct; but I would not have you imagine that I understand that in God himself, there are accidents, or that he could be known through his accidents.

Dix. I do not attribute to you such a dull understanding; and I know that it is one thing to say that the things extraneous to the divine nature are accidents, another thing to say that they are His accidents, and still another thing to say that they are, though, His accidents. I believe that you want to say that the effects of the divine activity exist in the last manner; which, although they are the substance of things, and even the natural substances themselves, nevertheless, are, as it were, the most remote accidents, whereby we touch the apprehensive cognition of the divine supernatural essence.

Theo. Well said.

Dix. Of the divine substance then—as well because it is infinite as because it is extremely remote from those effects which are at the farthest limit of the course of our discursive faculties—we can know nothing, except by the method of traces, as the Platonists say; of remote effects, as the Peripatetics say; of the raiment, as the Cabalists say; of the shoulders or back, as the Talmudists say; of the mirror, the shadow, and the enigma, as the Apocalyptics say.

Theo. Nay, even more; because we do not see perfectly this universe, whose substance and principle is with such difficulty comprehended, it follows that with far less grounds can we know the first principle and cause through its effect than Apelles can be known through the statues he has made; because we may see the latter entirely and examine them part by part, but not so, the vast and infinite effect of the divine omnipotence. Therefore, this comparison must be understood without any notion of proportionality.

Dix. So it is, and so I understand it.

Theo. It will be well, then, to abstain from speaking of so lofty a matter.

Dix. I consent to that, because it is sufficient, morally and theologically, to know the first principle in so far as the higher divinities have revealed it, and the prophets have declared it. Beyond this, not only whatever religion and theology you will, but also all advanced philosophy has declared it to be characteristic of a profane and turbulent spirit to rush in demanding reasons and definitions for those things which are above the sphere of our intelligence.

Theo. Good; but these do not deserve blame so much as those deserve praise who strive toward the knowledge of this principle and cause, and who come to know its grandeur as much as possible by allowing the eyes of well-regulated thoughts to scan those magnificent stars and those luminous bodies, which are so many inhabited worlds and great animals and excellent Gods; which seem to be, and are, innumerable worlds, not much unlike this one which contains us; which, since it is impossible that they have their existence in and of themselves, in view of the fact that they are composite and dissoluble (although not for that reason are they deserving of dissolution, as has been well said in the Timaeus), it is necessary that they know the principle and cause and, consequently, with the grandeur of their existence, their life and works, show and affirm, in infinite space, with innumerable voices the infinite excellence and majesty of their first principle and cause. Leaving then, as you say, these considerations, inasmuch as they are superior to all sense and intellect, we shall consider principle and cause, in so far as it is either nature itself, or in so far as it reveals itself to us in the extent and lap of that nature. Question me, then, in order, if you want me to answer you in order.

Dix. I will do so. But first, since you continually speak of cause and principle, I would like to know whether they are taken by you as synonymous names?

Theo. No.

Dix. Well, then, what difference is there between the one and the other term?

Theo. I answer that when we call God first principle and first cause, we mean one and the same thing from divergent points of view; when we speak of principles and causes in nature, we speak of different things from different points of view. We call God first principle, inasmuch as all things are after him, according to a certain order of before and after, either according to their nature, or according to their duration, or according to their worthiness. We call God first cause inasmuch as all things are distinct from him as the effect from the efficient [cause], the thing produced from the producer. And these two points of view are different, because not everything which is prior and more worthy is the cause of that which is posterior and less worthy, and not everything that is cause is prior and more worthy than that which is caused, as is very clear to him who considers carefully.

Dix. Then tell me, what difference is there between cause and principle in objects of nature?

Theo. Although at times one term is used in place of the other, nevertheless, properly speaking, not everything that is principle is a cause, because a point is the principle of a line, but it is not the cause of the line; the instant is the principle of activity, the "terminus a quo" is the principle of motion, the premises are the principle of the argument—but the former are not the cause of the latter. Therefore, principle is a more general term than cause.

Dix. Then, restricting these two terms to distinct and proper signification, in accordance with the custom of those who speak more correctly, I believe that you wish to say that the principle is that which intrinsically contributes to the constitution of things, and remains in the effect; as is said of matter and form which remain in their composite; or again, the elements of which the thing has been composed, and into which it is resolved (are principles). You call cause that which contributes to the production of things from without, and which has its being outside of the composition, as is the case with the efficient cause, and the end to which the thing produced is ordained.

Theo. Completely correct.

Dix. Since, then, we have reached an agreement regarding the difference between these things, I desire that you devote your attention first to the discussion of causes and then to the discussion of principles. And as to the causes, I should like first to know of the first efficient cause, of the formal cause, which you say is joined to the efficient, and then, of the final cause which is understood to be the mover of this.

Theo. The order which you propose pleases me very much. As to the efficient cause, I say that the universal physical efficient cause is the universal intellect, which is the first and principal faculty of the world soul, which [the soul] is the universal form of that [the world].

Dix. Your opinion seems to me to be not only in agreement with that of Empedocles, but more certain, more distinct and more explicit; moreover, in so far as I can see from the above, more profound. Therefore, you will be doing me a favor if you will explain the whole more minutely, beginning by telling me just what this universal intellect is.

Theo. The universal intellect is the most intimate, the most real, and the most proper faculty and partial power of the world soul. This is one and the same thing which fills the whole, illumines the universe and directs nature to produce its various species as is fitting, and has the same relation to the production of natural things as our intellect to the parallel production of rational concepts. This is called by the Pythagoreans the mover and arouser of the universe, as the poet has expressed,

> Infused through the members, mind vitalizes the whole mass and is mingled with the whole body.

This is called by the Platonists the world architect. This builder, they say, proceeds from the higher world, which is entirely one, to this sensible world, which is divided into many and in which not only love but also discord reigns, on account of the difference of its parts. This intellect, infusing and bringing something of its own into matter—restful and moveless in itself—produces everything. By the Magi this intelligence is called the most fecund in seeds, or the seed sower, since it is He who fills matter with all of its forms, and according to the type and condition of these actually shapes, forms, and arranges it [matter] in such admirable order, as cannot be attributed to chance, or to any principle which cannot distinguish and arrange. Orpheus calls this intellect the eye of the world because it sees all natural things, both from within and without, in order that all things may actually produce and maintain themselves in their proper symmetry, not only intrinsically, but also extrinsically. By Empedocles, it is called the distinguisher, since it never wearies of unfolding the confused forms within the bosom of matter or of calling forth the generation of one thing from the corruption of the other. Plotinus calls it the father and progenitor because it distributes the seeds throughout the field of nature, and is the proximate dispenser of forms. We call this intellect the inner artificer because it shapes matter and figures it from within, as from within the seed or the root it sends forth and unfolds the trunk, from within the trunk it sends forth the branches, from within the branches the formed twigs, and from within the twigs the buds—and within those it shapes, forms, weaves, as with nerves, leaves, blossoms and fruits; and from within, at certain times it recalls the sap from the blossoms and fruits to the twigs, from the twigs to the branches, from the branches to the trunk and from the trunk to the root. And it is similar with animals—its work proceeding first from the seed, and from the center of the heart, to the external members, and from these finally gathering back to the heart the unfolded powers, it acts as if again rolling together the unwound threads. Now if we believe that even inanimate works, such as we know how to produce with a certain order, imitatively working on the surface of matter, are not produced without reason and intellect, as when, cutting and carving a piece of wood, we bring forth the effigy of a horse, how much greater then must we believe is that molding intellect which, from the interior of the germinal matter, solidifies the bones, extends the cartilage, hollows the arteries, breathes out the pores, weaves the fibers, spreads out the nerves, and with such admirable mastery disposes the whole? How much greater I say an artificer is he who is not restricted to one part of the material world, but as a whole operates continually throughout the whole. There are three kinds of intelligences: the divine which is all things, the mundane which makes all things, and the other particular ones which become everything; because it is necessary that between the extremes this intermediary must be found, which is the true efficient cause, not only extrinsic but also intrinsic, of all natural things.

Dix. I should like to see you distinguish how you understand it to be extrinsic and intrinsic cause.

Theo. I call it extrinsic cause because as efficient it does not form a part of the things composed and the things produced. I call it intrinsic cause in so far as it does not operate around and outside of matter, but as has just been stated above. Hence, it is extrinsic cause with regard to its being which is distinct from the substance and essence of its effects, and because its being is not like that of things capable of generation and corruption, although it operates around those things; it is intrinsic cause with regard to the actuality of its workings.

Dix. It seems to me that you have spoken enough about the efficient cause. Now I would like to understand what kind of a thing is that which you take as formal cause, joined to the efficient cause; is it perhaps the ideal concept? Because every agent that works according to the rule of intelligence does not strive to produce effects unless in accordance with some image; and that image is not without the apprehension of something; and this is nothing else but the form of the thing which is to be produced. And, therefore, this intellect, which has the power to produce all species, and to send them forth with such beautiful construction from the potentiality of matter into actuality, must possess all of them in advance, after the manner of forms, without which forms the agent could not proceed to their production, just as it is impossible for the sculptor to execute diverse statues without first having inwardly imagined the diverse forms.

Theo. You understand this excellently, because I desire that two kinds of form should be considered—one which is the cause, not indeed the efficient cause, but that through which efficient cause works; the other is the principle, which is called forth from matter by the efficient cause.

Dix. The aim and the final cause which is pursued by the efficient cause is the perfection of the universe, which means that all the forms are actually existent in diverse parts of matter; the intellect takes such pleasure and delight in this end that it never tires of calling forth all sorts of forms from matter, as it appears that Empedocles would also wish it.

Theo. Entirely correct; and I add to this that just as this efficient cause is universal in the universe, and is special and particular in the parts and members of that, just so its form and its end.

Dix. Enough has been said about causes; let us proceed to the principles.

Theo. In order to arrive at the constitutive principles of things, I will first discuss form; for this is in some way the same as the aforementioned efficient cause because the intellect which is a power of the world soul has been called the proximate efficient cause of all natural things.

Dix. But how can the same subject be at one and the same time principle and cause of natural things? How can it have the role of an intrinsic part, and not of an extrinsic part?

Theo. I say that this is not unsuitable considering that the soul is within the body as the pilot is within the ship: which pilot, in so far as he shares the motion of the ship, is part of that; considered in so far as he governs and moves it, he is understood not as a part but as a distinct efficient cause. Just so the soul of the universe, in so far as it animates and informs, is an intrinsic and formal part of that [universe]; but in so far as it directs and governs, it is not a part: it has not the role of a principle, but of a cause. Aristotle, himself, admits this, who, although he denies that the soul has that relation to the body which the pilot has to the ship, however, considering it according to that power which understands and knows, he does not dare to call it an act or form of the body; but he understands it as an efficient cause, separate according to existence from matter, saying that that is a thing that comes from outside, according to its own subsistence, separated from the composite.

Dix. I approve of what you say, because if that existence separate from the body belongs to the intellectual power of our soul, and if this intellectual power has the role of an efficient cause, much more should the same thing be affirmed of the world soul; because Plotinus says, writing against the Gnostics, "that the world soul rules the universe with far greater ease than our soul rules our body." Beside there is a great difference in the way in which each one rules: the former, as if unbound, rules the world in such a way that it bends all that which it gets hold of; the former is not passive to other things; the former rises without hindrance to (superior) supernatural things; in giving life and perfection to the body, it does not itself receive any imperfection from that body; and, therefore, it is eternally joined to the same subject. The latter [human soul] is manifestly in a contrary condition. Since, then, according to your principles, the perfections which exist in inferior natures in a far higher degree should be attributed to and recognized in superior natures, we ought without a doubt confirm the distinction that you have made. This comes to be affirmed, not only in the world soul, but also in every star, since it being the case, as the aforementioned philosopher wishes, that they all have the power of contemplating God, the principles of all things, and the distribution of the orders of the universe; and he is not of the opinion that this happens through memory, reasoning, and consideration, because each of their works is an eternal work, and there is no activity which can be new to them, and, therefore, they do nothing which is not entirely fitting—perfect, with a certain and predetermined order—without activity of cogitation; as Aristotle shows this also, through the example of the perfect writer, and lyre player, when, in this, that nature does not reason and reflect, he does not wish it to be concluded that she works without intelligence and final end, because exquisite writers and musicians pay less attention to what they are doing, and yet do not err as do the inexpert and clumsy, who with more thinking and attending do their work less perfectly, and not without mistakes.

Theo. You understand me. Let us now proceed to the more particular. It seems to me that they detract from the divine goodness and from the excellence of this great animal and image of the first principle, who will not understand nor affirm that the world with all its members is animate; how should God be envious of his image, or how should the architect not love his own particular work, of whom Plato says that he takes pleasure in his work because of his own similitude which he sees reflected in that; and certainly, what more beautiful object than this universe could be presented to the eyes of deity? And since it consists of many parts, to which of these parts should more be attributed than to the formal principle? I shall leave for a better and more detailed discourse a thousand natural reasons beyond this topical and logical one.

Dix. I do not care to have you exert yourself in that direction, especially since there is no philosopher of any reputation, even among the Peripatetics, who does not hold that the world and its spheres are in some way animated. I would now like to know in what way you take it that this form makes its way into the matter of the universe.

Theo. It joins itself to it in such a manner that corporeal nature, which according to itself is not beautiful inasmuch as it is capable of it, participates in beauty, since there is no beauty which does not consist of some figure or form, and no form which has not been produced by the soul.

Dix. It seems to me that I am hearing something entirely new. You wish to hold perhaps that not only the form of the universe, but all forms of natural objects are souls?

Theo. Yes.

Dix. Are all things then animated?

Theo. Yes.

Dix. But who will grant you this?

Theo. But who with reason can say otherwise?

Dix. It is common sense that not all things are alive.

Theo. The sense most common is not the truest.

Dix. I easily believe that this can be defended. But the fact that a thing can be defended is not enough to make it true, inasmuch as it also must be proved.

Theo. This is not difficult. Are there not philosophers who say that the world is animated?

Dix. There are surely very many, and they are most authoritative.

Theo. Then why will not the same ones say that all the parts of the world are animated?

Dix. They surely do say that, but only of the principal parts, and those which are true parts of the world; especially since with no less reason they hold the soul to be as a whole throughout all the world, and in whatsoever part of it, than the soul of living things perceptible to us is present as a whole throughout the whole.

Theo. What things do you think, then, are not true parts of the world?

Dix. Those that are not what the Peripatetics call primal bodies: the earth, together with its waters and other parts, which, according to what you say, constitute the entire animal (organism); the moon; the sun; and other bodies. Besides these principal animals (organisms), there are those which are not primary parts of the universe, of which some are said to have a vegetative soul, some a sensitive soul, others the intellective soul.

Theo. Yet, if the soul since it is in the whole is also in the parts, why do you not hold that it is in the parts of the parts?

Dix. It may be, but in the parts of the parts of animate things.

Theo. Now what are these things which are not animate, or are not the parts of animate things?

Dix. Does it seem to you that we have so few of these things before our eyes? All the things which have no life.

Theo. And what are these things that have no life, or at least the vital principle?

Dix. To resolve this, then, do you hold that there is not anything that does not have a soul, and which does not have a vital principle?

Theo. This, in fine, is what I wish to hold.

Pol. Then a dead body has a soul? Then my shoes, my slippers, my boots, my spurs, my ring and my gloves will be animate? My coat and my mantle are animated?

Gerv. Yes sir, yes, Master Polyhymnius, why not? I well believe that your coat and mantle are well animated, when they have within them an animal such as you; the boots and spurs are animated when they contain the foot; the hat when it contains the head, which is not without a soul; and the stall is also animated when it contains the horse, the mule, your excellency. Do you understand it so, Theophilus? Does it not seem to you that I have understood it better than the *dominus magister*?

Pol. Cuium pecus? As if there are not to be found subtle asses; *etiam atque etiam?* How dare you, you trifler, you a, b, c, darian, compare yourself with a teacher and moderator of the school of Minerva, like me?

Gerv. Peace be with you, master and lord, I am the servant of your servants, and the footstool of your feet.

Pol. The Lord curse you, forever.

Dix. No quarreling; let us determine these things.

Pol. Then let Theophilus continue his teaching.

Theo. I shall do so. I say, then, that the table as a table is not animate, nor the garments, nor the leather as leather, nor the glass as glass— but as natural things and composites they have in themselves matter and form. Let a thing be as small and little as you wish, it has in itself some part of spiritual substance, which, if it finds a fitting subject, extends itself so as to become a plant, an animal, and receives the members of whatever body you wish, such as is usually said to be animated: because spirit is found in all things, and there is not the least corpuscle that does not contain in itself some such portion which spirit does not animate.

Pol. Therefore, whatever is, is an animal.

Theo. Not all things which have souls are called animate.

Dix. Then, at least, all things have life?

Theo. I concede that all things have in themselves a soul, and have life, according to substance and not according to actuality and operation, which alone is knowable to all the Peripatetics and all those who define life and soul according to certain gross principles.

Dix. You show me some plausible method in which the opinion of Anaxagoras may be upheld, who held that everything is in everything, because, since spirit, or soul, or universal form exists in all things, all can be produced from all.

Theo. I do not say plausible, but true; because that spirit is found in all things; those which are not animals are yet animate; they are not perceptible in their act as animals and living beings, yet they are perceptible with reference to the principle and certain first act of animality and life. And I cannot say more, because I wish to supersede the properties of many stones and gems, which, when they are broken and recut and placed in disordered pieces, have a certain virtue, in altering the spirit and engendering new affections and passions in the soul, not only in the body. And we know that such effects do not proceed, nor can they come from purely material quality, but they must necessarily be referred to a symbolic principle, vital and living; besides, we see this same thing sensibly in withered plants and roots, which, purifying and collecting humors, and altering the spirits, show necessarily the effects of life. I submit that not without cause the necromancers hope to produce many things by means of the bones of the dead, and they believe that those retain, if not the very same, yet a certain activity of life, which becomes useful in producing extraordinary results. On other occasions, I shall be able to discuss more at length concerning the mind, the spirit, the soul, and the life which penetrates all, is in all, and moves all matter, fills the womb of that matter and surpasses it rather than it is surpassed by it; for the spiritual substance cannot be overpowered by the material [substance] but rather contains it.

Dix. This appears to me to conform not only to the sense of Pythagoras, whose meaning the poet recites when he says:

> In the beginning the sky, the earth, and the fields of the waters
> Glistening out of the moon, and also the radiant sunlight
> All is inspired with life, and trembling through every member
> Mind vitalizes the mass, and the whole body is mingled.

but it also conforms to the sense of the wise king who says: "The spirit rules over and fills the earth, and that it is which contains all things."

And still another, speaking perhaps of the relations of form with matter and potency, says that that is dominated by the act and by the form.

Theo. If, then, spirit, mind, life is found in all things, and in various degrees fills all matter, it certainly follows that it is the true act and the true form of all things. The world soul, then, is the formal constitutive principle of the universe, and of that which is contained in it. I declare that if life is found in all things, this soul emerges as the form of all things—that which presides over matter, through everything; that which is master of composites, effects the composition and consistency of the parts. And, therefore, persistence belongs no less to such form than to matter. This I understand to be One in all things, which, however, according to the diversity of the dispositions of matter, and according to the power of the material principles, active and passive, comes to produce diverse configurations, and to effect different powers, sometimes showing the effect of life without sense, sometimes the effect of life and sense without intellect, and sometimes it appears that all the powers are suppressed and repressed either by weakness or by other conditions of matter. Thus, while this form changes place and condition, it is impossible that it be annulled; because the spiritual substance is not less subsistent than the material. Then only external forms change themselves and are even annulled because they are not things, but of things; they are not substances, but of substances—accidents and circumstances of substances.

Pol. Non etia sed entium.

Dix. Obviously, if any substantial thing could be annulled, the world would become empty.

Theo. We have, then, an intrinsic principle, formal, eternal, subsistent, incomparably better than that which the Sophists have imagined, who play with accidents, ignorant of the substance of things, and who are led to posit corruptible substances because they call chiefly, primarily, and principally, that substance which results from composition; the latter is not other than an accident, containing in itself no stability and truth, and resolves itself into nothing. They call that the true man which results from composition; they call that the true soul which is either the perfection and act of a living body, or at least a thing which results from a certain symmetry of complexion and members; whence it is not so astounding that they do so much and so greatly fear death and dissolution; as those on whom the loss of their being is immanent. Against this madness nature cries out in a loud voice, assuring us that neither bodies nor souls should fear death, because matter as much as form are the most constant principles:

O race, atremble with fear, with the icy terror of dying,
Wherefore dread ye the Styx, vain names, and the forms of shadows
Idle subjects for parts, and dangers of the world that exist not?
Whether the funeral pile shall consume our bodies with fire.
Or old age wasting away, think not that we can suffer evil
Souls are not subjects to death, but former dwellings abandoned
Rise to shelter eternal, where they may inhabit forever.
Thus do all things suffer change, but nothing ever shall perish.

Dix. That appears to me to conform to the saying of Solomon, judged the wisest of men among the Hebrews:

> The thing which has been, it is that which it shall be. And that which is done, is that which shall be done. And there is no new thing under the sun.

So that this form which you presume is not existent in and adherent to the matter, according to its existence, does not depend upon the body and upon matter to the end that it subsists?

Theo. Thus it is. And besides, I do not determine whether all form is accompanied by matter, just as of matter, I surely say, that of that, there is no part that is, in fact, without form, unless you understand form abstractly as Aristotle does, for the latter never wearies of dividing conceptually that which is indivisible according to its nature and truth.

Dix. Do you not affirm that there may be some other form besides this eternal companion of matter?

Theo. Of course, and a form more natural still than is the material form, of which we shall reason later. For the present, note this distinction of form; there is, namely, one sort of primal form which informs, is extended, and is dependent; and this, since it informs everything, is in all; and since it is extended, it communicates the perfection of the whole to the parts; and because it is dependent and has no activity from itself, it communicates the activity of the whole to the parts—similarly, the name and existence. Such is the material form, like that of fire; because every part of fire warms, it is all called fire and is fire. Secondly, there is another kind of form which informs and is dependent, but is not extended; and such a form because it perfects and activates the whole, is in the whole and in every part of the whole; because it is not extended, it results that it does not attribute the act of the whole to its parts; because it is dependent, it communicates the activity of the whole to the parts. And such is the vegetative and the sensitive soul, because no part of the animal is animal; and, nevertheless, each part lives and feels.

Third, there is another kind of form, which actuates and perfects the whole but is not extended, nor is it dependent as regards its operations. The latter, because it actuates and perfects, is in the whole, and in each and every part; because it is not extended, it does not attribute the perfection of the whole to the parts; because it is not dependent, it does not communicate the activity (of the whole to the parts). Such is the soul, inasmuch as it can exercise intellectual power, and is called intellective soul; the latter does not form any part of man that can be called man, or is man, or can be said to know.

Of these three kinds of form, the first is material, which cannot be understood nor can it exist without matter; the other two kinds (which in fine concur in one according to their substance and their being, and are distinguished according to the method which we have spoken of above) denominate that formal principle, which is distinct from the material principle.

Dix. I understand.

Theo. Further than this, I want you to take notice that although, according to common speech, we say that there are five levels or kinds of form—that is, the elemental, the mixed, the vegetative, the sensitive, and the intellective—we do not, therefore, understand this according to the usual [vulgar] intention; because this distinction is valid according to the operations which appear and proceed from subjects, not according to that ground of the primary and fundamental existence of that form and spiritual life which itself fills all things, and not according to the same manner.

Dix. I understand. Inasmuch as this form which you posit as principle is a subsistent form, it constitutes a perfect species, is of its own genus, and is not part of a species like that peripatetic form.

Theo. Thus it is.

Dix. The distinction of forms in matter is not according to accidental dispositions which depend upon the material form.

Theo. True.

Dix. Whence also this separate form does not come to be multiplied according to number, because every numerical multiplication depends on matter.

Theo. Yes.

Dix. Besides, though invariable in itself, it is variable through particular subjects and the diversity of matter. And such form, although in the subject it makes the part differ from the whole, (yet itself) does not differ in the part and in the whole; although one reason suits it as subsistent in and by itself, and another in so far as it is the activity and perfection of some subject, and still another in regard to a subject with dispositions of some kind, and another with those of another.

Theo. Absolutely so.

Dix. This form is not to be understood as accidental, nor as similar to the accidental, nor as mixed with matter, nor as inherent in that, but rather as existing in, associated, and assistant.

Theo. That is what I say.

Dix. Besides, this form is defined and determined by matter; because having in itself the facility to constitute particular things of innumerable species, it contracts itself to constituting one individual; and, on the other hand, the potentiality of indeterminate matter, which can receive whatsoever form you will, comes to be determined into a species; so that the one is the cause of the definition and determination of the other.

Theo. Very good.

Dix. Then, you approve, in some way, the opinion of Anaxagoras, who calls the particular forms of nature latent; and in a way that of Plato, who deduces them from Ideas; and in a way that of Empedocles, who makes them proceed from intelligence; and in some way that of Aristotle, who makes them, so to speak, issue from the potentiality of matter.

Theo. Yes, because, as we have said, where there is a form, there is in a certain way everything; and where there is soul, spirit, life, there is everything; for the intellect is the shaper of the ideal species, and if it does not bring the forms forth from matter, it none the less does not go begging for them outside of matter, because this spirit fills the whole.

Pol. I want to know in what way the form is everywhere the whole soul of the world, if it is indivisible? It is necessary, then, that it is very big, even of infinite dimensions, if you call the world infinite.

Gerv. There is good reason for it to be large, as also a preacher at Grandazzo in Sicily said of our Lord: where as a sign that He is present throughout the world, he ordered a crucifix as big as the church itself, in the similitude of God, the Father; who has the Empyrean heavens for a canopy, the starry heavens for his throne, and has such long legs that they reach down to the earth, which serves him as a footstool. To him came a certain peasant, who asked him—Reverend father, now how many ells of cloth would it take to make his breeches? And another said that all the peas and beans of Melazzo and Nicosia would not be sufficient to fill his stomach. See to it, then, that this world soul is not made in the same fashion.

Theo. I do not know how to respond to your doubt, Gervasius, but perhaps I can respond to that of Master Polyhymnius. I can however, to satisfy both of you, give you an analogy, because I wish you to take away with you some fruits of our reasoning and discourse. Briefly, you ought to know that the world soul, and the Divinity, are not omnipresent through all and in every part in the manner in which material things could be there, because this is impossible to any kind of body, and to any kind of spirit; but in a way which is not easy to explain to you, if not in this way. You ought to take notice that if the world soul and the universal form are said to be everywhere, we do not mean corporeally and dimensionally, because they are not such; and just so they cannot be in any part; but they are spiritually present in all, as for example (albeit a rough one), you can imagine a voice which is as a whole throughout a whole room, and in each part of that room; because as a whole, it is heard throughout all; just as these words which I am saying are themselves heard as a whole by all, even if there were a thousand present, and my voice, could it reach throughout the world, would be everywhere as a whole. I say to you then, Master Polyhymnius, that the soul is not indivisible like the point, but in some way like the voice. And I respond to you, Gervasius, that the Divinity is not everywhere in the sense that the God of Grandazzo is in the whole of the chapel because, although he is present throughout the church, yet all of him is not present everywhere, but he has his head in one part, his feet in another, his arms and his chest in still other parts. But that other is in its entirety in whatsoever part, just as my voice is heard entirely in each part of this room.

Pol. I perceived that perfectly.

Gerv. I have heard your voice at least.

Dix. I well believe it of the voice, but as regards the discourse, I think that it has gone in at one ear and out through the other.

Gerv. I think that it has not even gone in; it is late, and the clock which I have in my stomach has struck supper time.

Pol. This is what it is to have your brains in a platter.

Dix. Enough then. Tomorrow perhaps we shall meet to discuss the material principle.

Theo. Either I will expect you, or you may expect me here.

End of the Second Dialogue

THIRD DIALOGUE

Gerv. The hour has arrived, and they have not yet come! While I have no other thing to do that attracts me, I wish to gain for myself the pleasure of listening to them. Apart from what I can learn of the other chess movements of philosophy, I can besides have an enjoyable pastime, learning about the fancies that dance in the marvellous brain of that pedant, Polyhymnius. The latter first declares that he wishes to judge who speaks well, who discusses better, who commits incongruities and contrarieties in philosophy; and afterwards, when his time comes to speak his part, he does not know what he should put forward, and releases a mass of his windy pedantry, a salad of proverbial sentences of Latin or Greek that never has any pertinence to what the others have said; whence, any blind man can see without too much difficulty what a fool he is despite all his Latin, while it is obvious that the others are wise men in their simple vernacular. Yet, by my faith, here he is already! As he walks towards here, it is apparent that his very way of walking is the result of his Latin. Welcome the "dominus magister."

Pol. I put very little value on this "magister"; since in this absurd and topsy-turvy age such a title has come to be attributed to those of my station and alike to every barber, artisan, and sow cutter; therefore, it is good council for us: *Nolite vocari Rabbi!*

Gerv. How do you wish me then to address you? Would "most reverend" please you?

Pol. That is befitting the cleric and the Priest.

Gerv. Do you wish then the title, "most illustrious"?

Pol. *Cedant arma togae!* That title is befitting as well for those of the knightly station, as for those of courtly standing.

Gerv. Imperial majesty?

Pol. *Quae Caesaris, Caesari.*

Gerv. Accept then the title, "Lord"! Take the title of thunderer, of father of the Gods. Let us come to facts: why are you so late?

Pol. I think the others are also occupied in some other business, just as I am; and so as not to leave this day pass without writing a line, I have given myself up to the contemplation of that kind of globe that has been commonly called a map of the world.

Gerv. What have you got to do with the map of the world?

Pol. I am contemplating the parts of the earth, the zones, the provinces, and the regions; all of which I have taken in and gone over in my mind, and many have I covered with my steps also.

Gerv. I should like you to investigate yourself somewhat; because it seems to me that this is more important, and I think that you do not occupy yourself enough with that.

Pol. Absit verbo invidia; because in this other way I come to know myself more efficiently.

Gerv. And how will you persuade me of that?

Pol. From the contemplation of the macrocosmos we can easily, after having made necessary deductions, arrive by means of comparison to a knowledge of the microcosmos, the small parts of which correspond to the large parts of the macrocosmos.

Gerv. And thus we should find in you then, the moon, Mercury, and the other stars, France, Spain, Italy, England, Calicut and the other countries?

Pol. Quidni? Per quandam analogiam!

Gerv. Per quandam analogiam. I believe that you are a great monarch; but when you were a woman, then I would ask you if there may be a place in you to shelter an infant, or to preserve any of the other plants of which Diogenes spoke.

Pol. Ah, ah, a good joke! But such questions are not in accordance with the station of a wise and highly educated man.

Gerv. If I were an educated man or if I held myself to be wise, I would not come here to learn with you.

Pol. Not you, but I do not come here in order to learn, because it is my office to teach; and it is my interest to judge those who wish to teach; therefore, I come here with another intention than that for which you ought to come here, to whom the role of beginner, novice, and pupil is befitting.

Gerv. What then is your purpose?

Pol. In order to judge, I say.

Gerv. Truly, it is far more fitting for one of your station to judge the sciences and doctrines, because you are the only ones to whom the liberality of the stars and the munificence of fate has conceded the ability to distill the sweet sap from the words.

Pol. And consequently from the thoughts, too, because they are connected to the words.

Gerv. As the body to the soul.

Pol. So that it is only when the words are rightly understood that the sense is fully grasped; hence from the knowledge of languages—and in this I am more versed than any other person in this city, and I consider myself just as learned as anyone else who runs a school—and so, what I want to say is that out of the knowledge of languages proceeds the knowledge of any science whatsoever.

Gerv. Then all those who understand the Italian language, will understand the philosophy of the Nolan?

Pol. Of course, but it is also necessary that one possess dexterity and judgment.

Gerv. For some time, it has been my idea that this dexterity was the principal thing; because one, who does not know the Greek, can understand the entire meaning of Aristotle and recognize many errors in that philosophy. Clearly, one sees that this idolatry which surrounded the authority of that philosopher, especially in natural science, has been abolished by those who understand the teachings of that other school; and one, such as Paracelsus, who knows nothing of Greek, of Arabic, and perhaps nothing of Latin, can have a better knowledge of the nature of drugs and the art of medicine than Galen, Avicenna, and all the others who speak in the Roman language. Philosophies and laws do not fail through the lack of interpreters of words but rather through the lack of such as are profound in their thoughts.

Pol. So you place me in the category of the stupid crowd?

Gerv. God forbid. Because I know that in virtue of the knowledge and the study of languages—and this is a rare and extraordinary thing—not only you, but all the likes of you, are very well qualified to pass judgment on those doctrines, after having defeated the opinions of those who have championed them.

Pol. Because you speak the truth, I can very easily persuade myself that you do not speak without good reason; and inasmuch as it is not difficult for you, there should be no serious difficulty in explaining it.

Gerv. I will do this, subjecting myself always to the judgment of your prudence and learning: it is a common proverb that those who are outside of the play, understand better than those who are within the play; as those who are in the theatre, can better judge of the acts, than those persons who are in the scene (itself); and music can better be tested by one who is not in the orchestra or in the choir; similarly, is this observable in cards, chess, fencing, and others of this kind. And so is it also with you, gentlemen pedants, that through being excluded and outside of every act of science and philosophy, and through not having and never having had any familiarity with Aristotle, Plato, and the like, can better judge and condemn them with your grammatical sufficiency and arrogant reliance on your natural capacity. The Nolan, who finds himself in the same theatre, in such familiarity and intimacy with them, easily combats them after having recognized their innermost and most profound opinions. I declare, then, that you, through being outside every profession of the honored and ingenious men, can better judge them.

Pol. I am unable to answer this impudent man on the spot. *Vox faucibus haesit.*

Gerv. However, are the men of your type more presumptuous than those who have both their feet in the question; and, therefore, I assure you that worthily you usurp the office of approving this, disapproving that, glossing the other, making here a concordance and collation, there an appendix.

Pol. This most ignorant of all men wishes to infer from the fact that I am learned in humane letters, that I am ignorant in philosophy.

Gerv. My most learned gentleman, Polyhymnius; I wish to say that if you had all the languages that there are, as our preachers say, seventy-two . . . .

Pol. Cum dimidia.

Gerv. . . . . It would not follow from this that you may be able to judge philosophy, but more than this, that you could not help being the most clumsy animal that exists in human form; and, on the other hand, it is no impediment to one who scarcely knows one of these languages, even a bastard language, to be the most wise and the most learned man in the world. Yet consider, what success two such men have acquired: one, a Frenchman, an archpedant, who has written discussions on the liberal arts and who has written observations against Aristotle; the other, an Italian, a truer excrement of pedantry who has soiled many pages with his "Discussiones Peripateticae." Everyone easily sees that the first most eloquently shows what little knowledge and understanding he has; the second, simply speaking, shows that he has much of the beast and the ass. Of the first we can at least say that he understood Aristotle, but he understood him badly; and if he had understood him well, he would perhaps have had the ingenuity to make honored war against him, as has the most judicious Telesio, of Cosenza. Of the second, we cannot say that he has understood him either well or badly; but he has read and reread him, sewed, ripped apart, and compared him with a thousand other Greek authors, friends and enemies of the aforementioned; and finally, he has made a very enormous work, not only without any profit, but also with the greatest harm; in this way, he who wishes to see into what folly and presumptuous vanity, a pedantic character can precipitate and sink, should look into this book only, before it loses its seed. But here are Theophilus and Dixon.

Pol. Adeste felices, domini! Your presence is the cause that my wrathfulness will not issue into lightning judgments against the vain statements that this garrulous charlatan has made.

Gerv. And you have made short my enjoyment in sporting about the majesty of this most reverend owl.

Dix. Everything goes well, so long as you do not get angry.

Gerv. Everything that I say, I say in jest, because I love my teacher.

Pol. I too, though rough, am not seriously enraged because I do not hate Gervasius.

Dix. Good, allow me then to discuss with Theophilus again.

Theo. Democritus, then, also the Epicureans, say that that which is not corporeal is nothing; and, consequently, they hold matter alone to be the substance of things; and the same to be the divine nature; and the same thing was said by a certain Arab, called Avicebron, as he shows in a book entitled "Fons vitae." The same philosophers, along with the Cyrenaics, the Cynics, and the Stoics, hold that the forms are nothing else but certain accidental dispositions of matter. And for a long time, I myself have been an adherent to this conception, only because it has a foundation more corresponding to nature than the views of Aristotle; but after having more maturely considered, and after having regarded more things, we find that it is necessary to recognize in nature two kinds of substance: one of which is form, the other of which is matter; because it is necessary that there be a most substantial act, in which is the active potency of everything; and also a potency and a substratum in which there is equally large passive potency of all; in the former is the power to make, in the latter the power to be made.

Dix. To every thinking person it must be manifestly clear that it is impossible that the former can always make everything without there always being that which can become everything. How can the world soul, that is, every form which is indivisible, grant shape, without the substratum of dimensions or quantity, that is, matter? And matter—how can that become informed? Perhaps through itself? It is apparent that we can say that matter becomes formed through itself, if we consider the universal formed body to be matter and want to name it matter; as we would call an animal with all its faculties, matter distinguishing it, not from the form, but only from its efficient cause.

Theo. No one can prevent you from making use of the term matter in your manner; because the word has different meanings in many schools, reason has many different meanings. But this manner of consideration that you speak of will, I know, not stand up well, except to a mechanic or a doctor who is interested in practice, as, for example, to him who divides the entire universe into mercury, salt, and sulphur; this manner of speaking shows not so much a divine ingenuity in medicine as much as it shows a fool who wishes to call himself a philosopher; whose purpose is not to come to that distinction of principles which is made physically through the separation that proceeds in virtue of fire; but also to that distinction of principles to which no material efficient cause arrives because the soul, which is inseparable from sulphur, from mercury, and from salt, is a formal principle; the latter is not the subject for material qualities, but is entirely the sovereign of matter; it is not touched at all by the work of the chemists, whose division ends in the three aforementioned things, and who know another kind of soul than that of the world, and which we ought rightfully to define more closely.

Dix. You speak most excellently; and this consideration pleases me very much because I see that there are many people of restricted judgment and little insight who do not distinguish the causes of nature taken absolutely, in accordance with the entire circuit of their being, as philosophers do, but from them as taken in a limited and restricted sense; the first mode is for the doctors as such, worthless and excessive, the second is defective and insufficient to the philosophers as such.

Theo. You have touched upon the very point for which Paracelsus, who has treated philosophy medically, is to be praised, and for which Galen is to be censured, inasmuch as the latter has treated medicine philosophically; the latter, by making such a disgusting mixture and such a confused picture produces in the end but a poor doctor, and a very much confused philosopher. But this must be said with some reserve because I have not had the leisure to examine all the aspects of the work of that man.

Gerv. Pardon, Theophilus, do me this favor first—since I am not so experienced in philosophy—expound for me what you understand under the name of matter, and what is truly matter in natural things.

Theo. All those who wish to distinguish matter and consider it in itself, separate from form, recur to the analogy of Art. So it is with the Pythagoreans, the Platonists, and the Peripatetics. Take, for example, an art like that of carpentry: it has wood as a subject for all of its forms and for all of its work; just so is it with the blacksmith whose subject is iron, as the subject of the tailor is cloth. All of these arts bring into a particular matter diverse pictures, arrangements, and figures, none of which is proper and natural to matter. Therefore, nature, to which art is similar, needs to have for its operations a matter; because it is not possible that there be an agent which, when it wishes to make something, does not have that out of which it can make it; or likewise, if it wishes to work, does not have that on which to work. There is then a kind of substratum from which, with which, and in which, nature effects its operations and its work; and which is by nature endowed with so many forms that it presents for our consideration such a variety of species. And just as wood in itself has no artificial form, but can have all of them through the operations of the carpenter, in a similar way, matter, of which we speak, has no natural form by itself, and in its nature, but can have all forms through the operations of the active agent, the principle of nature. This natural matter is not, therefore, as perceptible as artificial matter is because the matter of nature has absolutely no form; but the matter of art is a thing already formed by nature, since art can only work on the surface of things formed by nature, as is the case with wood, iron, stone, wool, and others of this kind; but nature, so to speak, works from within its subject or matter, which throughout is formless. Therefore, the subjects of art are manifold, but the subject of nature is one; because the former being diversely formed by nature, are different and variegated; the latter, being formless, is entirely indifferent, since all difference and diversity stem from the form.

Gerv. So that the things formed by nature are the matter of art, and one thing alone that is formless is the matter of nature?

Theo. So it is.

Gerv. Is it then possible that just as we can see and know clearly the substrata of the arts, we can similarly know the substratum of nature?

Theo. Without a doubt, but with diverse principles of cognition; for just as we do not know colors and sounds by means of the same sense, similarly, we cannot see by means of the same eye the substratum of the arts and the substratum of nature.

Gerv. You wish to say, then, that we see the former with the sensible eyes, and the latter with the eyes of reason.

Theo. Completely right.

Gerv. May it please you then to develop this argument.

Theo. Willingly. That same relation and reference which the form of art has to its matter is likewise, according to due analogy, had by the form of nature to its matter. Just as in art, the forms varying to infinity (if this were possible), there always remains under all those forms the one same matter; as, for example, after the form of the tree, there is the form of the trunk; so then that of the board, that of the table, bench, stool, chest, the comb, and so forth; and throughout all the wood remains the same; so analogously, is it in nature, where although the forms vary themselves to infinity, the one succeeding the other, the matter remains ever the same.

Gerv. And how would you corroborate this comparison?

Theo. Do you not see that that which was seed becomes a herb, that that which was herb becomes corn, that that which was corn becomes bread—from bread, chyle; from chyle, blood; and from this, seed, from this Embryo, man, corpse, earth, stone, and something else, and so on, taking in all of the natural forms?

Gerv. I see this very easily.

Theo. It is necessary, therefore, that there be one same thing which in itself is not stone, not earth, not corpse, not man, not embryo, not blood, or anything else; but that which, after it was blood, became embryo, receiving the embryonic being; and after it was embryo received the human being, becoming man (himself); just so that which is formed by nature, which is the subject of art; for example, that which was a tree becomes a table, and received the being table; that which was table, receiving the being door, becomes a door.

Gerv. I have understood this very well. But, it seems to me, this subject of nature cannot be corporeal, nor of a certain quality; because this subject which is escaping now under one form and natural being, now under another form and being, does not show itself corporeally, as the wood or the stone, which make themselves seen as that which they are while they serve as material or subject, under whatever form they may appear.

Theo. Completely correct.

Gerv. What shall I then do, when it shall come to pass that in discussing this thought with some obstinate person, he does not wish to believe that there is one matter under all the forms of nature, just as there is one under all the formations of each art? Because the latter which can be seen with the eyes cannot be denied, but the former which can be seen with reason only can be denied.

Theo. Send them away, or do not answer him!

Gerv. But what if he will be steady in his demand for evidence, and if he will be a respectable person who can send me away rather than I can impudently send him away, and who will take it as an insult when I do not answer him?

Theo. What would you do if a blind demigod, worthy of whatever honor and respect, will be bold, steady, and obstinate in wishing to have knowledge and evidence of colors, or even of the exterior forms of natural things; as, for example, what is the form of the tree? What is the form of the mountains? Of the stars? Further, what is the form of the statue, of the robes? And in short of all the other artificial things, which to those who see are so manifest?

Gerv. I would answer him that if he had eyes he would not demand such evidence, but would be able to see them by himself; but being blind, it is also impossible for anyone else to show them to him.

Theo. Similarly, will you be able to answer those—that they, if they had intellect, would not demand any other evidence, but would be able to see them by themselves.

Gerv. Some would be ashamed by this answer, others would consider it too cynical.

Theo. Then you will say to him more secretly thus: My most illustrious gentleman, or, your sacred majesty, since some things cannot become evident except through the hands and the sense of touch, and others with the ears, and still others only with the tongue, and others with the eyes—so this matter of natural things cannot become evident except through the intellect.

Gerv. Perhaps that one, understanding the blow as not being so dark or so secret, will say to me: You are the one that has no intellect; I have more than all those of your standing.

Theo. You would not then believe him more than a blind man who would say to you that you are blind, and that he sees more than all those who think they see, as you do.

Dix. Enough has been said in demonstrating more evidently than I have ever heard what the name matter signifies, and what one must understand as matter in natural things. So Timaeus the Pythagorean, who, from the transmutation of one element to another, teaches one to find the matter which is hidden, and which cannot be known, except by means of certain analogy. Where the form of earth was, he says, afterwards appears the form of water; and it cannot be said that one form receives the other, because one contrary does not accept or receive the other; that is, the dry does not receive the wet, nor the dryness the wetness; but the dryness is expelled from a third itself and the wetness is introduced and this third thing is the substratum of both contrary qualities—itself not contrary to anything. Then, too, since one must not think that the earth has disappeared into nothing, one must judge that some thing which was in the earth has subsisted, and is in the water; which thing, for the same reason, when the water shall be transmuted into air (by virtue of the heat that comes to extenuate it into gas and vapor), will remain, and will be in the air.

Theo. From this it can be concluded, in spite of the Peripatetics, that nothing is annihilated or loses its being, but only its exterior and material accidental form is lost. Therefore, the matter, and the substantial form of anything in nature—which is the soul—cannot be dissolved, or annihilated, completely losing their being. To be sure, then, all the substantial forms of the Peripatetics and of others like them, cannot be of this character, since they consist of nothing else than in a certain complexion and order of accidents: and for them, everything that they will know how to declare outside of their prime matter, is nothing else than accidents, complexions, habit of quality, principle of definition, quiddity. Whence some cowlish, subtle, metaphysicians among them, wishing to excuse rather than to accuse the insufficiency of their God, Aristotle, have discovered humanity, bovinity, oliveness, as specific substantial forms; this humanity, as Socratiety, this bovinity, this horseness, are the individual substances. They have done all this in order to give us a substantial form which merits the name of substance, just as matter has the name and being of substance; but they have at no time profited anything through this; because if you ask them rightfully, what the substantial being of Socrates consists of, they will answer—in Socratiety. If you ask further, what do you understand by Socratiety, they will respond—the proper form and the proper matter of Socrates. Let us leave aside this substance which is matter and ask: what is the substance as form? Some will answer: its soul. You ask further—what kind of a thing is this soul? If they say: an entelechy and perfection of a body which has the capacity to live, please consider that this is but an accident. If they say: it is the principle of life, sense, vegetation, and intellect, please consider that, although this principle is, fundamentally considered, as we consider it, a substance, nevertheless, our opponents do not consider it except as accident; because to be the principle of this or that is not the same as being the absolute and substantial reason, but an accidental and relative reason of that which is derived; in the same way, that which expresses what I do or can do is not expressing my being and substance but rather that which expresses what I am (being and substance) as myself absolutely considered. You see, therefore, how they treat this substantial form which is the soul; which although it has been recognized by them as substance, they have never really named it or considered it as substance. This conclusion you can see more manifestly when you ask them: what constitutes the substantial form of an inanimate thing, for example, wood? Those of them who are more subtle will feign: in woodness. Now take away that matter which is common to iron, wood, and stone, and now say, what remains of the substantial form of iron? They will never answer you with anything but accidents; these, however, belong to the principles of individuation and cause particularity; for the matter cannot be in any other way limited to particularity, except through a form; and this form, because it is the constituent principle of a substance, should be substantial; but afterwards they will not be able to show it in reality except as an accident; and finally, when they will have done everything that they can, they have a substantial form, yes, but not one present in nature, but a logical one; and so finally it comes to pass that some logical intention comes to be assumed as the principle of natural things.

Dix. Has not Aristotle realized this?

Theo. I believe that he had most certainly realized this, but he could not help himself; therefore, he declared the last differences to be unknown and unnamable.

Dix. Therein, it seems to me, has he confessed his ignorance; and, therefore, I would yet judge it better to embrace those principles of philosophy which in this important question do not plead ignorance, such as Pythagoras, Empedocles, and your Nolan, whose opinion we touched on yesterday.

Theo. This is the opinion the Nolan wishes to express: that there is one intellect that gives being to everything; this is called by the Pythagoreans and Timaeus the giver of forms; that there is one soul and formal principle that becomes and informs everything; this is called by those aforementioned ones the fountain of forms; that there is one matter out of which everything is produced and formed; this called by all the receptacle of forms.

Dix. This doctrine pleases me very much because it shows nowhere any defect. And verily is it a necessary thing that just as we can posit a constant and eternal material principle, so also we should posit in a similar way a formal principle. We see all the forms in nature cease in matter and again appear in matter; whence there seems to be no other thing in reality except matter, constant, firm, and eternal, and worthy of estimation as principle. Besides that, the forms have no being without matter in which they are generated and corrupted and out of whose bosom they spring and into whose bosom they are taken back. Therefore, matter, which always remains fecund and the same, must rightfully be given the prerogative of being recognized as the only substantial principle; and as that which is, and always remains; all the forms together must be taken merely as various dispositions of matter, which come and go and cease and renew themselves; and, therefore, not all can have the reputation of principle. Therefore, there are to be found some who, having well considered the reason of natural forms, as far as it has been possible to know from Aristotle and other similar authors, have finally come to the conclusion that they are nothing but circumstances and accidents of matter; and, therefore, the prerogative of act and perfection must be referred to matter and not to things, of which we can truly say they are not substance, nor nature, but things of substance and of nature; but this, they say, is matter, which according to them is a necessary principle, eternal and divine, just as it is to that Moor Avicebron, who calls it God who is in all things.

Theo. They have been led to this error because they know no other form except the accidental form; and this Moor, although he had accepted the substantial form of the Peripatetic doctrine in which he was nurtured, withal, considering it as a corruptible thing, not only mutable concerning matter, and as that which is produced and does not produce, established and not establishing, rejected and not rejecting, depreciated it and held it vile in comparison with stable, eternal, producing, maternal matter. And surely this happens to all those who do not know that which we know.

Dix. This has been very well considered, but it is time that we leave this digression and return to our problem. We now know how to differentiate matter from form, as well from the accidental form (be that as you wish) as from the substantial form; what remains now for us to look into is its nature and reality. But first I would like to know whether, in virtue of the great union that this world soul and this universal form has with matter, that other mode of philosophizing ought to be accepted, of those who do not separate the activity from the concept of matter and consider this [matter] as something divine and not as nude and formless in such a manner that she would not give herself form and dress herself (with various forms).

Theo. Not easily, because nothing works absolutely on itself; for there is always some distinction between that which is the agent, and that which is produced, or concerning which the action and the operation takes place; for that reason it is good to distinguish in the body of nature, the matter from the soul, and in this to distinguish the general from the specific kinds. Therefore, we say, that there are three things in this body: first, the universal intellect inherent in things; second, the vivifying soul of all; third, the substratum. But because of this we shall not deny the name of philosopher to those who conceive in accordance with their kind of philosophy this formed body, or as we wish to say, this rational animal; and forthwith begin to accept in a way as first principles the limbs of this body, such as water, air, earth, fire; or ethereal region and the stars; or spirit and body; or vacuum and plenum (understanding the vacuum not as Aristotle grasped it); or in some other adequate way. It does not appear to me that this philosophy is to be rejected, especially when, no matter what foundation they presuppose, or form of edifice that they propose, it brings into effect the perfection of speculative science and the knowledge of natural things, as it has truly been done by many ancient philosophers. For it shows an extremely ambitious person, a presumptuous, vain, and invidious brain to want to persuade others that there is but one way to investigate and come to the knowledge of nature; only a stupid fool and a person without reason convinces himself of that. Therefore, the most constant and steady way, the most contemplative and clear way, and the mode of consideration which is highest ought always to be preferred, honored, and sought after; but that other mode is not to be censured that is not without good fruit, though they may not be from the same tree.

Dix. Do you approve, then, the study of diverse philosophies?

Theo. Quite so, for those who have the time and the ingenuity; for the others I approve the study of the best, when the Gods wish that they find it out.

Dix. Therefore, I am certain that you do not approve all philosophies, but only the good and the best ones.

Theo. So it is; so also in the different orders of medicine. I do not reprove him who makes use of magic, through applications of roots, the affixing of stones, and the murmuring of incantations, if the rigor of the theologians allows me to speak as a mere philosopher of nature. I approve that which is done physically (through nature) and is executed through the apothecary's prescriptions, with which the cholera, blood, the phlegm, and melancholia can be fought or gotten rid of. I have nothing against the others who proceed (chemically) by way of chemistry, who abstract the quintessence, and through mediation of fire, cause from out of the whole composition the mercury to rise, the salt to subside, and the sulphur to become luminous or be dissolved. But as regards medicine, I do not wish to establish, among so many good methods, which is the best; because the epileptic, over whom the physician and the chemist have lost time, will, and not without good reason, if he is cured by the magician, approve the latter more than any other doctor. In an analogous manner we must argue about all the other kinds; no one of which will seem to be less good than the other, if so much one as the other realizes the end that (she proposes to herself) is proposed. Now in a particular case, that doctor who cures me will be better than the others who kill and torment me.

Gerv. Wherefore does it happen, then, that there are so many enemies among these sects of doctors?

Theo. From avarice, jealousy, from ambition and ignorance. Usually, they hardly understand the proper method of cure, let alone that of the others. Moreover, the greater part of them, not being able to raise themselves to honor and profit, with proper virtue, try to get preference by blaming the others and by depreciating that which they cannot acquire. But of these, the best and the truest is he who is not only a physicist but also a chemist and a mathematician. And to return to our point, among the kinds of philosophy, that is best which more properly and more highly effects the perfection of the human intellect, and which more properly corresponds to the truth of nature, and as far as possible goes hand in hand with the latter (I mean on an ordered natural way and through the knowledge of change); not through animal instinct as do the beasts, and these others who are like them; not through the inspiration of good or bad demons, as do the prophets; not through melancholy enthusiasm, as the poets and other contemplators—or ordering laws or reforming customs, or curing, or even knowing and living a life more blessed and divine. You see, then, how there is no kind of philosophy that is steadily ordered by regulated sentiment which does not contain in itself some good property that is not contained in the others. The same thing I understand to apply to medicine, which is derived from such principles, that presuppose a not imperfect habit of philosophy; as the activity of the foot or the hand presupposes that of the eye. Therefore, it is said that no one who has not made a good ending in philosophy can make a good beginning in medicine.

Dix. I am very much pleased with you, and I praise you highly for it. That you are not so uncultured as Aristotle, and also not so slanderous and ambitious as he; who wanted all the opinions and methods of all the other philosophers to be disparaged.

Theo. Though of so many philosophers that there are, I know of none that is founded more on imagination and more removed from nature than he; and if he sometimes says some excellent things, you should know that they do not depend on his principles, but are always propositions taken from some other philosophers; and we can see many such divine things in the book on Generation, in the Meteors, in the books on Animals and Plants.

Dix. Returning, therefore, to our problem: do you hold that matter can be defined diversely without error and without involving oneself in contradiction?

Theo. True, for the same subject can be judged by diverse senses, and can present itself in diverse ways. Moreover, as has been pointed out, the consideration of a thing can be grasped under diverse conceptions. The Epicureans have said many good things, though they have never elevated themselves above material quality. Heraclitus has made known many fine things, though he has not taken himself above the soul; Anaxagoras is not lacking in making progress into nature, because not only within the latter but perhaps even outside and above it, he conceives of an intellect which is called God by Plato, Socrates, Trismegistus, and our theologians. Therefore, no less well can he proceed to discover the secrets of nature who starts from the experimental reason of the simples (so called by them) than he who starts from theoretical reasoning; and among these, no less he who starts from complexions than he who starts from humors; and this no more than he who descends from sensible elements or he who starts higher up, from the absolute elements, or from one matter, which is the highest and most distinct principle of all. Indeed, sometimes he who takes the longer way does not make the more successful trip, especially when his end is not so much theory and contemplation as operation. Concerning the manner of philosophizing, then, it will be no less convenient to have the forms evolve from *implicato* than to distinguish them as coming out of chaos, or to distribute them as from an ideal fountain, or to bring them into act from potency, or to report them as coming from a womb, or to bring them out to the light, as from a dark and blind abyss; because every foundation is good when it is approved by the edifice; every seed is welcome when the trees and the fruits are desirable.

Dix. So to come now to our aim, please declare the distinct doctrine of this principle.

Theo. Certainly; the principle which has been called matter can be considered in two ways: first, as a potency, second, as a substratum. In the first meaning, taken as a potency, there is no thing in which, in a certain way and according to proper reason, it can not be found; the Pythagoreans, the Platonists, the Stoics, and others have placed it in the intelligible world as well as in the sensible world. But we, not understanding it in exactly the same way as they understood it, but in a higher and more explicit sense, speak in this sense of potency or possibility. Potency is usually divided into active potency, by means of which the subject of it can operate, and into passive potency, by means of which it either can be, or can receive, or can have, or can be the subject of the efficient cause in any manner. Not reasoning about the active potency at present, I say that the potency which signifies the passive mode, though it is not always passive, can be considered either relatively or absolutely. And, therefore, there is no thing that can be said to be, which cannot be said to have the capacity to be. And this in such a manner corresponds to the active potency, that one is not without the other in any way; whence, if there always has been the potency to make, to produce, to create, there always has been the potency to be made, to be produced, to be created; because the one potency implies the other; I want to say that one being posited, it necessarily posits the other. This potency does not indicate weakness in that of which it is said, but rather confirms its virtue and efficacy; and at the end is found to be all one and indeed the same thing as active potency; and hence there is no philosopher, nor theologian, who doubts that it can be attributed to the first supernatural principle. Because the absolute possibility through which the things which are in act can be is not before that actuality, nor after that; moreover, the capacity to be is together with the being in act, and does not precede that; because if that which can be could make itself, it would be before it was made. Now contemplate the highest and best principle, which is all that it can be; it itself would not be all if it could not be all; in it, therefore, the act and the potency is the same thing. This is not the case with other things which, though they are that which they can be, could, however, not be, and certainly could be something else or in a different way from what they are now; for no other thing is all that it can be. The man is that which he can be, but he is not all that he can be. The stone is not all that it can be because it is not chalk, it is not a vase, it is not dust, and it is not a herb. That which is all that it can be is one, which comprehends and contains in its being all being. It is all that is and can be whatever other thing that is and can be. Every other thing is not so; for here the potency is not equal to the act because the act is not absolute but limited; moreover, the potency is always limited to one act because it never has more than one specific and particular being; and when it nevertheless refers itself to every form and act, it does so through certain dispositions and through a certain order of succession of one being after another. Every potency and act, then, which in the principle is enfolded, unified, and one, is in other things unfolded, dispersed, and manifold. The universe, which is the grand simulacrum, image, and infinite nature, is also all that it can be, through the same species and principal members, and being inclusive of all matter, to which nothing is added, and from which nothing is lacking, and of all and one form; but it is yet not all that it can be through the same differences, modes, properties, and individuals; therefore, the universe is only a shadow of the first act and the first potency; and in it the potency and act are not absolutely the same because no one of its parts is all that it can be. Otherwise the universe, in that specific mode which we have declared, is all that it can be; and this in a developed, dispersed, and distinct way. Its principle, on the contrary, is all that it can be, unifiedly and indifferently, because it is all as a whole, and the same [in all] simply, without distinction and difference.

Dix. What would you say of death, corruption, vices, defects, and monsters? Do you wish to make it your meaning that these, too, have a place in that which is all that it can be, and is in act all that which it is in potency?

Theo. These things are not act and potency, but defect and impotency; these are found in things that are enfolded, because they are not all that they can be, and are compelled to be what they can be; whence, not being able to be so many things at once and simultaneously, they lose one being in order to have the other; and sometimes they confuse one being with another; and sometimes they are reduced, imperfect, and mutilated through the incompossibility of this being and that, and through the preoccupation of matter in this and that being. Nevertheless, returning to our subject, the first absolute principle is sublimity and greatness, and is such sublimity and greatness in such a way that it is all that it can be. It is not great in such a way that its greatness can be made greater or lesser, nor can it be divided, as can every other kind of greatness which is not all that it can be; hence, it is the greatest and the smallest greatness, infinite, indivisible, and of every measure. It is not the greatest, because it is the smallest; it is not the smallest because it is likewise the greatest; it is above any equality because it is all that which it can be. This that I say concerning its greatness, you should likewise understand of whatever can be said; for it is similarly the goodness which is all the goodness that can be; it is the beauty that is all the beauty that can be; and there is no other beauty, outside of this one, that is all that it can be. One is that which is all and can be all, absolutely. In natural things, moreover, we do not see anything which is other than that which it is in act, and accordingly it is that which it can be, in virtue of having a specific kind of actuality; nevertheless, in this unique specific being, there is never all that that one particular thing can be. Here is the sun: she is not all that the sun can be; she is not in all places, where the sun can be, because when she is east of the earth, she is not in the west, nor in the north, nor in the south, nor in any other aspect. Therefore, when we want to show the manner in which God is the sun, we will say (because He is all that He can be) that He is simultaneously in the east, west, south and north, and in whatsoever point of the earth's globe; whence, if of this sun (either through her own revolution or that of the earth), we want to understand whether she moves and changes her place (we will say) that she is not actually in any point, without the capacity to be in all the others; and, therefore, has the ability to be there: and, therefore, if she is all that she can be, and possesses all that she is able to possess, she will simultaneously be throughout all in all; and she is the most moveable and the fastest in such a way that she is also the most stable and immobile. Therefore, we find among the divine sayings, that she has been called stable in eternity, and the fastest thing that runs from end to end; for that is understood as immobile which in the one same instant leaves the eastern point and returns to the eastern point, and for that reason it is not less seen in the east than in the west, and in every other point of its circuit. Therefore, we find there is no more reason for saying that she goes and returns or has gone and returned from point to that than from any other of the infinity of points to the same. Whence she will come to be entirely and always in all the circle, and in whatever parts of that; and, consequently, every individual point of the ecleptic contains the entire diameter of the sun. And so an indivisible comes to contain the divisible; this does not happen through natural possibility but through supernatural possibility; I meant to say this when I presupposed that the sun was that which is in actuality everything that it can be. The absolute potency itself is not only that which the sun can be, but that which everything is, and that which everything can be: potency of all the potencies, act of all the acts, life of all lives, soul of all souls, being of all being; whence, the highly extolled utterance of revelation: "He who is, sends me"; "He who is, speaks also." Therefore, that which is elsewhere contrary and opposite is in Him one and the same, and everything is in Him the same; and so you may argue about the differences of time and duration, as we did about the differences of actuality and possibility. Thus there is in Him nothing new and nothing old; whence this has been well uttered by revelation: the first and the last.

Dix. This absolute act, which is the same as absolute potency, cannot be comprehended by the intellect, except in a negative way—that is, it cannot be apprehended, neither in so far as it can be everything, nor as it is everything: because the intellect, when it wishes to understand, must form the intelligible species, assimilate it, measure itself with it, and make itself equal to that: but this is here impossible, because the intellect never is such that it cannot be greater; the other, however, inasmuch as it is in every sense and from all sides immeasurable, cannot be greater than it is. There is, therefore, no eye capable of approaching and having access to this highest light and this deepest abyss.

Theo. The coincidence of this act with the absolute potency has been very clearly described by the divine spirit where he says: "Yea, the darkness hideth not from there, but the night shineth as the day: the darkness and the light are both alike to Thee." To conclude, then, you see how great the excellency of that potency is, which if it pleases you to call it the essence of matter (into which those vulgar philosophers have not penetrated), you can treat in a higher meaning than Plato has done in his *Republic* and in his *Timaeus*, without subtracting from the divinity. These, through having raised the essence of matter to too high a plane, have caused a scandal to some theologians. This has resulted either because the former have not understood it well or because having been brought up in the views of Aristotle, they always take the meaning of matter as it is as the subject of natural things; they do not consider that, for the others, matter is such as to be something common to the intelligible and sensible world, as they say, taking its meaning in an analogous equivocation. Therefore, before they are condemned, the opinions ought to have been very well examined, and the terms ought to have been as much distinguished as the notions; especially in view of the fact that, though all sometimes agree to a common concept of matter, they differ afterwards in their particular notion of it. And what now concerns our subject, it is impossible, prescinding from the name matter, that there is to be found a theologian, be he deceitful, or however malevolent, who can impute to me impiety for what I say and understand by the coincidence of potency and act, taking the one and the other term absolutely. I would like now to conclude, in such proportion as it is permitted, to say about that image of that act and potency; for it is in specific activity all that which it is in its specific possibility in so far as the universe, in this way, is all that which it can be—(be it as it will, in relation to individual activity and individual potency); and thus it comes to have a potency which is not absolute from act, a soul not absolute from the animated; I do not mean the composed, but the simple; whence it is understood that there is in the universe a first principle, more distinctly material and formal, which can be inferred from the similitude of the aforesaid as absolute potency and act. Whence it is not made difficult and risky to accept, finally, that the whole in its substance is one, as perhaps Parmenides understood it, whom Aristotle has treated very ignobly.

Dix. You wish, then, that, although in descending through this scale of nature one comes upon two substances, one spiritual, the other corporeal, they are in the end reduced to one being and root.

Theo. If it appears to you that this can be tolerated by those who do not pursue the problem further.

Dix. Easily, when you do not raise yourself above the terms of nature.

Theo. This has already been done. If we do not have the same conception and method of speaking about the divinity as the common one, yet our conception is particular—yet in no way contrary nor alien to the other, and yet perhaps clearer and more explicit—and is in accordance with reason that does not go above the head of our discourse and from which I did not promise you I would abstain.

Dix. Enough has been said about the material principle, according to the point of view of possibility or potency; will you please, therefore, tomorrow, proceed to the consideration of the same, from the point of view of substratum.

Theo. That I will do.

Gerv. Goodbye.

Pol. May the omens be favorable for us.

END OF THE THIRD DIALOGUE

FOURTH DIALOGUE

Pol. Et os vulvae nunquam dicit: sufficit: id est, scilicet, videlicet, utpote, quod est dictu: matter (which is expressed by these words) *recipiendis formis nunquam expletur.* And since there is no one else in this lyceum, or rather antilyceum, thus I alone—I say alone, that is, really less alone than anyone else—shall take a stroll and I shall hold dialogue with myself. Matter, then, for the prince of the Peripatetics and the preceptor of the highest spirit of the great Macedonian, not less than for the divine Plato and the others has been called (now) chaos, (now) hyle, (now) Silva, (now) mass, (now) potency, (now) aptitude, (now) admixture of privation, (now) cause of evil, (now) ordered to evil, (now) non-being in itself, (now) in itself unknowable, (now) knowable through analogy to form, (now) *tabula rasa*, (now) indescribable, (now) subject, (now) substratum, (now) substerniculum, (now) a field, (now) infinite, (now) indetermination, (now) *prope nihil*, (now) neither *quid*, nor *quale*, nor *quantum*—thus, after having tired myself with various and many different nomenclatures: in order to better define this essence, I have come to entitle it "woman" from all pertinent attributes; finally, I say, to comprise all in one word, it is called woman by those who consider the matter more carefully. And by Hercules, the senators of the kingdom of Pallas have wanted, and not without some unusual reason, to place these two things—matter and woman—on an equal footing: since they have been brought to that rage and madness by the experience had with the rigor of them—(here a rhetorical figure occurs to me). They are a chaos or irrationality, a hyle of wickedness, a forest of ribaldry, a mass of filthiness, an aptitude to every perdition (I have here another rhetorical figure, called by some, Complexio!): where was (already in potentiality—not only far but also near) the destruction of Troy? In a woman. Who was the instrument of destruction for Samson's strength, I mean, of that hero who, with the jaw of an ass that he found, became the invincible victor over the Philistines? A woman! Who mastered, at Capua, the violence and force of the great general and perpetual enemy of the Roman republic, Hannibal? A woman! (Now comes an *Exclamatio!*) Tell me, O prophet of the Cithern, the cause of your fragility? "Because my mother has conceived me in sin." How did it happen (Oh, our great ancient father) that you, being a gardener in paradise, and a caretaker of the tree of life, came to such a malevolent state that you threw yourself, together with the entire germ of humanity, into the deep abyss of perdition? "The woman, whom he gave me; she, she has deceived me." Without doubt, form itself does not sin, and error proceeds from no form, except in so far as it is united to matter. In this way, the form, which is signified by the male, having been placed in familiarity with matter, and having come into composition or unification with that, responds to ***natura naturans*** with these words, or better with this sentence: "The woman, whom you gave me—that is, the matter which you have given me as a wife—she has deceived me"—this is the cause of all my sin. Think, how the egregrious philosophers and the discrete anatomists of the secrets of nature have not found a more adequate manner for more fully presenting the nature of matter than to confront us with this proposition which shows that the state of natural things is to matter as the economical, political, and civil state is to the feminine sex. Open, open your eyes, etc.—Oh, I see that colossus of laziness, Gervasius, who interrupts the thread of my skilful speech. I do not know whether he has heard me, but it does not matter.

Gerv. Greetings, O master, most excellent of all teachers!

Pol. If you do not—as you usually do—wish to mock me—greetings to you too.

Gerv. I would like to know why you were walking around alone and brooding?

Pol. While I was studying in my small museum, I came across (in Aristotle) the passage at the end of the first book of the *Physics*, where he wants to make clear what prime matter is, and in which he takes as his mirror [example] the feminine sex: the sex, I mean, stubborn, fragile, inconstant, soft, petty, infamous, ignoble, rude, despicable, negligent, unworthy, reprobate, sinister, vituperous, frigid, deformed, vacuous, vain, indiscrete, insane, perfidious, indolent, offensive, filthy, ingratious, cut off, mutilated, imperfect, inchoate, insufficient, precise, amputated, attenuated, rancorous, caterpillar-like, turbulent, pestiferous, morbid, dead—

> To us by God and nature sent
> Us to plague as burden and punishment.

Gerv. I know that you say this more to practice the art of oratory and to demonstrate how eloquent and copious you are than that you have such thoughts that you express through these words. For it is a very ordinary thing for you humanists, who call yourselves professors of liberal arts, that when you find yourselves full of those concepts which you cannot retain, you do not seek to unload them elsewhere than on the poor women; just as when any other grudge bothers you, you vent it upon the first delinquent among your students. But guard yourselves, you Orphei, from the furious indignation of the Thracian women.

Pol. I am Polyhymnius, not Orpheus.

Gerv. Then do you not censure the women truly?

Pol. I speak truly and I do not think otherwise than I speak; because I do not, as the sophists are wont, make it a profession to demonstrate that white is black.

Gerv. Why do you dye your beard then?

Pol. But I speak freely and I say that a man without a woman is like one of the intelligences (spirits); and he is a hero, a demigod, who has not married.

Gerv. And he is like an oyster, a mushroom, and a truffle.

Pol. Wherefore, the lyric poet has divinely said: "Believe, Pisones, the bachelor's life is the best." And if you wish to know the reason, listen to the philosopher Secundus: "Woman," he says, "is the impediment of quietude, a continuous infirmity, a daily war, a prison for life, a storm in the house, the shipwreck of a man." That man from Biscay well confirmed this who, having been made impatient and angry by the horrible fortune and fury of the sea, turned to the sea and with a scowling and angry face cried: O sea, sea, would that I could marry you—wishing to infer by this that woman is the tempest of tempests. Hence, Protagoras to the question—why he had given his daughter to his enemy—answered that he could do no more harm to his enemy than to give him a wife. Further, I will not be refuted by a good Frenchman, to whom, like all the others, when they suffered the dangerous tempest of the sea, having been commanded by Cicala, the master of the ship, to throw the heaviest burden into the ocean, threw his wife in first.

Gerv. You do not refer, on the other hand, to so many other examples of those men who have judged themselves very fortunate through their wives; and among them, without going any further, we have under this very roof the Monsieur de Mauvissier; he has one, who is not only gifted with unusual physical beauty, as cloak and covering of her soul, but further, with the triple virtues of discrete judgment, prudent modesty, and honest courtesy; she has tied the soul of her consort with an indissoluble knot, and is capable of captivating everyone who knows her. And what would you say about their generous daughter who is scarcely six years old; from her speech you cannot judge whether she is from Italy, France, or England; and her ability with musical instruments has made it difficult to decide whether she is a corporeal or incorporeal substance; and because of the mature excellence of her manners, one cannot know whether she has dropped from the sky or stems from the earth. And everyone must see that just as in the formation of her beautiful body the blood of both parents has concurred, so also in the fabric of her extraordinary spirit have been fused the virtues of the heroic soul of these same two.

Pol. A "Rara avis"—as Maria de Bochetel; a "Rara avis"—as Maria de Castelnau.

Gerv. This "rareness" that you say of women can also be said of men.

Pol. Finally, to return to our subject, woman is no more than a matter. If you do not know what woman is, because you do not know what matter is, study the Peripatetics a little, who by teaching you what matter is will teach you what woman is.

Gerv. I see plainly that since you have a peripatetic brain, you understood little or nothing of what Theophilus said yesterday about the essence and potency of matter.

Pol. Of the rest be that as you will; I want only to censure the appetite of the one and the other, which appetite is the cause of everything evil, of every passion, of every defect, of every ruin and corruption. Don't you think that if matter would content itself with its present form, no passion or alteration would be master over us—we would not die, and we would be incorruptible and eternal?

Gerv. And if it had contented itself with the form that it had fifty years ago, what would you say? Would you be Polyhymnius now? If it had been held in that form that it had forty years ago, would you be so adulterous, I mean to say so adult, perfect, and learned? Thus, as you so rightly show, all those forms have ceded their place to this; thus also it is the will of nature that orders the universe, that causes all forms to cede their places to others. And certainly it is a greater dignity for this, our substance, that it has the power to become anything, receiving all forms—than to be partial, keeping only one form. In this way, within its possibilities, it has a similarity to Him who is all in all.

Pol. You are beginning to become learned, leaving your ordinary nature. Apply then now this same thing, if you can, by way of similarity, to show the dignity that there is in woman.

Gerv. I will do that with consummate ease. But here is Theophilus.

Pol. And Dixon.—Another time, then. Enough for now.

Theo. We have already seen, have we not, that the Peripatetics, as also the Platonists, divide substance, by differentiating it into corporeal and incorporeal substance? And just as these differences are reduced to the potency of the same genus, so it is necessary that the forms be of two classes. Some are transcendent, that is, superior to genus, and are called principles, such as entity, unity, one, thing, something, and others of similar nature; the others are of a certain genus and distinct from another genus, as, for example, substantiality and accidentality. The forms of the first class do not make any distinction in matter and do not constitute distinct potencies of it, but as universal terms they comprehend all the corporeal, as well as the incorporeal substances, signifying that most universal, most common, and one potency of both. After this, what hinders us, says Avicebron, "That just as before we admit the matter of accidental forms, which is the composite, we recognize the matter of the substantial form, which is of that composite, so also, before we admit the matter which is contracted to be under corporeal forms, we recognize a potency that is distinguishable through the form of corporeal and incorporeal, dissoluble and indissoluble nature?" Besides, if all that which is, beginning from the highest and most supreme being, has a certain order and constitutes a subordination, a scale in which one rises from composed things to the simple, from these to the most simple and absolute, through corresponding means that combine and participate in the nature of the one and the other extreme, and are in their proper essence neutral, there is no order in which there is not a certain participation; there is no participation where there is not found a certain combination; there is no combination without a certain participation. It is necessary, therefore, that there be a principle of subsistent existence, of all [subsistent] existent things. Add to this that the same reason cannot help presupposing something indistinct before anything susceptible of distinction—I speak of the things that are; for I do not understand being and non-being to have a real distinction, but a verbal and nominal one only. This indistinct thing is a common essence, to which is joined the difference and the distinct form. And surely, it cannot be denied that just as everything sensible presupposes the substratum of the sensible, so also everything intelligible presupposes the substratum of intelligibility. It is necessary, then, that there be something that corresponds to the common concept of one and the other substratum, because every essence is necessarily founded on some existence, except that first essence that is identical with its existence, because its potency is the same as its act, and it is everything that it can be, as we have pointed out yesterday. Moreover, if the matter—according to the same opponents—is not body and precedes, according to its nature, the corporeal being, what is it that can make it so strange and heterogeneous from the substances called incorporeal? And there are not lacking Peripatetics who say that just as in corporeal substances there is something formal and divine, so also in divine substances it is fitting that there be something material, to the end that the inferior things be conformed to the superior, and the order of the former depends on the order of the latter. And the theologians, though some of them may be nourished on Aristotelian doctrine, must not be hostile against me in this, if they concede that they are more indebted to Scripture than to philosophy and natural reason. Do not adore me, said one of the angels to the patriarch Jacob, because I am your brother. Now, then, if this angel who speaks is an intellectual substance, as they themselves understand it, and affirms with his speech that that man and he himself agree in the reality of the same substratum—apart from whatever formal difference—it results that the philosophers have for proof an oracle of the theologians.

Dix. I know that you say this with reverence, because you know that it is not befitting to you to borrow reasons from such texts that are outside of our jurisdiction.

Theo. You speak rightly and truly; but I have not meant to allege that as reason and confirmation, but to evade scruples as much as I can; I fear no less to appear, than to be, opposed to theology.

Dix. Natural reasons will always be admitted by discreet theologians, inasmuch as they will discuss as long as they do not establish anything contrary to divine authority, but subject themselves to that.

Theo. Such are my reasons, and so will they ever be.

Dix. Good, proceed then.

Theo. Plotinus, also, in his book on matter, says that "if in the intelligible world there is a multitude and a plurality of species, it is necessary that there be something common, apart from the property and difference of each one of them. That which is common has the function of matter, that which is proper and makes for distinction, that of form." He adds that, "if this sensible world is the imitation of that intelligible world, the composition of this is the imitation of the composition of that. Besides, that world, if it lacks diversity, does not possess order; if it lacks order, it does not have beauty and ornament; all this concerns matter." Hence, the superior world must not only be considered as totally indivisible, but also in some of its aspects divisible and distinct; its division and distinction cannot be conceived without some matter that is its substratum. And although I say that all that multitude [multiplicity] coincides in one indivisible being, and is outside of every dimension, I will call matter that in which all those forms are united. This being, before being conceived as various and manifold, was in a uniform concept; and before being in a formed concept, was in a formless concept.

Dix. In what you have just explained briefly, you have brought forward many good reasons to conclude that there is only one matter, and one potency through which everything that is, is in act, and which with no less reason belongs to incorporeal as to corporeal substances, since the former no less have their existence through the capacity to be than the latter have their being through the power to exist, and since, moreover, through other powerful reasons, as is evident to him who considers and comprehends them—as you have demonstrated. Withal, if yet not so much to perfect this doctrine as to clarify it, I should like you to explain in some other manner how in the most excellent things—the incorporeals—the formless and the indeterminate can be found; how can the same matter be spoken of there, and how must that not equally be called body which results from the addition of form and act; how, where there is no mutation, no generation, nor any corruption, can you assume matter, which never has been assumed [posited] for any other end; how could you say that intelligible nature is simple, and affirm at the same time that there is in it matter and act. I do not formulate these questions for myself, because the truth of this doctrine is manifest to me, but perhaps for others, who may be more obstinate and harder to convince, as, for example, the Master Polyhymnius and Gervasius.

Pol. I accept.

Gerv. I accept and I am grateful to you, Dixon, because you have considered the necessities of those who haven't the courage to question—as conforms with the courtesy of the transalpine tables, where he who is sitting at the table in the second place is not allowed to extend his fingers outside of his proper square and plate, but he must wait until the food is placed into his hands in order that he may not take a bite which he has not paid with a "thank you."

Theo. I will say, in resolution, that just as man, in accordance with his proper nature of man, is different from the lion, according to the proper nature of the lion—but according to the common nature of animal, of corporeal substance and other similar things, they are indifferent and the same thing—so, similarly, according to its proper essence, the matter of corporeal things is different from that of incorporeal things. Everything, therefore, that you mention of its being the constitutive cause of corporeal nature, of its being the substratum of all kinds of transmutations, and part of compositions, belongs to that matter through its proper essence; because this same matter—I wish to say this more clearly—this same matter—that which it can be made or can be, or is made, it is in virtue of the dimensions and extension of the subject, and those qualities which have their existence in quantity; and this is called corporeal substance, and presupposes corporeal matter; or, it is indeed made, if indeed it has being newly and is without those dimensions, extension, and qualities; and this is called incorporeal substance, and similarly presupposes the said matter. Thus, to an active potency (common) as much to corporeal things as to incorporeal things, or to a being as much corporeal as incorporeal, there corresponds a passive potency, as much corporeal as incorporeal, and a capacity to be, as much corporeal as incorporeal. If therefore, we want to speak of composition as much in the one as in the other nature, we must understand it in this double sense; and must consider that it denotes in eternal things a matter which is always under one act, and in variable things it always contains now one, now another act: in the former, the matter has all that it can have, and is all that it can be, at once, always, and together; in the latter, on the other hand, it has it not at once, but at different times, and in a certain order of succession.

Dix. Some, although they concede matter in incorporeal things, understand it in a very different sense.

Theo. There may be whatever diversity according to the proper essence of each one—in virtue of which [the] one descends to corporeal being and the other does not, the one receives sensible qualities and the other does not—and it seems that there cannot be a concept common to that matter which is contrary to quantity and to being subject of those qualities which have their being in dimensions, and to that nature which is not contrary to either; nevertheless, they are one and the same, and in which, as we have often remarked, the entire difference depends on the fact that one is contracted to corporeal being and the other is contracted to incorporeal being. In a like manner, in the animal being, everything sensitive is one; but, contracting that genus to certain species, it is contrary to man to be a lion, and to this animal to be that other one. And I add to this, if it pleases you, because you will say to me that that which never is, must rather be judged impossible and unnatural [rather] than natural; and, therefore, since that matter is never found with dimensions, the corporeality must be judged unnatural to it; and this being so, it is not probable that the one and the other may have the same nature, before the one is understood to be contracted to be corporeal; I add, I say, that we can not less attribute to that matter the necessity of all dimensional acts—than—as you would desire to attribute to it their impossibility. That matter, through being actually all that it can be, has all the measurements, has all the species of figures and dimensions, and because it has all, has none of them; because for that which is so many different things it is necessary that it be no one of those particular things. It is proper for that which is all to exclude all particular being.

Dix. Do you wish to say, then, that matter is act? Do you wish also to say that matter in incorporeal things coincides with act?

Theo. As the capacity to be coincides with being.

Dix. It is, then, not different from form?

Theo. Not in absolute potency and absolute act. Which is, therefore, the highest grade of purity, simplicity, indivisibility, and unity, because it is absolutely all; for, if it had certain dimensions, certain being, certain figure, certain property, and certain difference, it would not be absolute, and would not be all.

Dix. Everything, then, which comprises any genus is indivisible?

Theo. Thus it is; for the form which comprises all qualities is not any one of those; that which has all figures has no one of those; that which has all sensible being is not, therefore, sensible. More highly indivisible is that which has all natural being; still higher is that which has all intellectual being; highest of all is that which has all the being that can be.

Dix. Do you wish to say that there is similar to this scale of being a scale of capacity to be? And do you wish to say that as the formal concept ascends, so does the material concept ascend?

Theo. That is true.

Dix. You take this definition of matter and potency very deeply and very highly.

Theo. True.

Dix. But this truth will not be able to be understood by all, because it is very difficult to grasp the way in which it can have all the kinds of dimensions, and none of them, and can have all the formal being and no one formed being.

Theo. Do you understand how this can be?

Dix. I think so, because I understand very well that the act, because it is all, does not have to be any particular thing.

Pol. I understand that too.

Theo. Then you will be able to understand that, if we wished to establish measurability for the concept of matter, such a concept would not be contrary to any kind of matter; but one matter will differ from the other only because the one is absolutely free from dimensions and the other is contracted to dimensions. As being absolute, it is above all, and comprises them all; as being contracted, it is comprised by some, and is under some.

Dix. You rightly say that the matter in itself has no determinate dimensions, and for that reason is understood as indivisible, and as receiving the dimensions by virtue of the form that it receives. It has some dimensions under the human form, others under the form of a horse, others under the form of an olive tree, others under a myrtle plant; for that reason, before it exists under any of those forms, it has in potentiality all those dimensions, just as it has the potency to receive all those forms.

Pol. Yet they say, on that account, that it has no dimensions.

Dix. And we say that it has none, in order to have all.

Gerv. Why do you wish so that it includes all, rather than that it excludes all?

Dix. Because it does not come to receive the dimensions from without, but sends them out from and casts them out as from her womb.

Theo. Very well said. Besides, it is a customary way of speaking for the Peripatetics, too, who all say that the activity of dimensions, and all the forms, issue from and proceed out of the potency of matter. This Averroes partly understood, who although he was an Arab and ignorant of the Greek language, understood more of the Peripatetic doctrine than any Greek that we have read; and he would have understood much more if he had not been so addicted to his God, Aristotle.• He says that the matter comprises in its essence the indeterminate dimensions; he wishes with this to hint that those dimensions come to be determined sometimes with this figure and these dimensions, sometimes with such and such others, according to the change in natural forms. In this sense, it is seen that matter sends the forms out from itself, and does not receive them from without. This in part Plotinus, the prince of Plato's sect, understood also; the latter, making the differences between the matter of superior and inferior things, says that the former is everything at the same time, and possessing all, it does not have to change itself into anything; but the latter, with certain changes in the parts, becomes everything, and according to different times, becomes different things; therefore, it is always under diversity, alteration, and movement. Therefore, the former matter is never formless, as the latter is not also, although each in a different manner; the former, in the instant of eternity; the latter, in the instants of time; the former, at the same time; the latter, successively; the former, as unfolding; the latter, as enfolding; the former, as many; the latter, as one; the former, through each and every thing; the latter, as all and everything.

Dix. So, you wish to infer, then, that not only according to your principles but also according to the principles of the other kinds of philosophy, matter is not that "prope nihil," that pure and bare potency without act, power, or perfection.

Theo. So it is; I speak of it as deprived of forms and without them, not in the manner in which the ice lacks warmth, and darkness is deprived of light, but in the manner in which the pregnant is without its progeny, which she sends forth and obtains from herself; and as the earth in this hemisphere is without light at night, the (same) light which she has the power to recover by her own revolution.

Dix. Here, then, even in inferior things, if not throughout, there is a coincidence of act and potency in many cases.

Theo. I leave this open to your judgment.

Dix. And if this potency from below became, finally, one with the potency from above, what would happen?

Theo. You can judge for yourself. You can raise yourself to the concept—not, I say, of the greatest and highest principle, because that is excluded from our consideration—but to the concept of the soul of the world, inasmuch as it is the act and the potency of all—omnipresent in all. And thence, finally, since there are innumerable beings, everything is one; and the knowledge of this unity is the object and term of all philosophies and all natural contemplations, leaving out in its own limits that contemplation which ascends above nature, which taken to him who does not believe is impossible and nothing.

Dix. True; because you ascend to that through the supernatural light, not the natural light.

Theo. This (supernatural light) is lacking to those who consider that everything is a body, either simple, as the ether, or composed as the stars and celestial things; and they do not look for the divinity outside of the infinite world and infinite things, but in these and within that.

Dix. It is in this alone, it seems to me, that the believing theologian differs from the philosopher.

Theo. I believe so too; I believe that you have understood what I wanted to say.

Dix. Very well, I think: in such a way that I infer from what you have said that although we do not allow that matter to be elevated above natural things, and although we support ourselves on the common definition, which the more common philosophy brings to it, yet we will find that it (matter) retains a better prerogative, than that (common definition) recognizes; the latter definition finally accords it nothing else but the concept of being a substratum of forms, of receptive potency of natural forms, without name, or definition, or some determination, because it lacks all actuality. This seemed difficult to admit to some of those men of the cowl who, wishing to excuse rather than accuse this doctrine, say that matter has only an entitative act—that is to say—which differs from that which simply is not, from that which does not have any being in nature, as occurs with some chimera or feigned thing; because this matter, in the end, has a being, and this is sufficient for it, without having quality or dignity which depends on actuality—which is a nothing. But then you should ask Aristotle: Why do you pretend, O prince of the Peripatetics, that matter is nothing, through not having some kind of act; rather than all through having all the acts, although they may be entirely confused or mixed, as you want to say? Are you not that one who, always speaking of the new being of the forms in matter, or of the generation of things, declare that the forms proceed and come out of the interior of matter; and are you not the one who has never been heard to say that through the workings of the efficient cause they come from without, but that the efficient cause brings them forth from within? And omit that the efficient cause of these things which you denominate by the common name, nature, is also conceived by you as an internal principle and not as an external one—as is the case with artificial things. It seems to me that you should correspondingly say that matter does not possess in itself some form or act, if it receives them from outside; but then I think it is fitting to say that it has all, if it is said to take all out of its own bosom. Are you not the one, then, who, when not constrained by reason—yet carried by the use of language—in defining matter, prefer to call it "that thing from which every natural species is produced"; and you have never said that it may be "that in which things are made," as it would be fitting to say if the acts did not come out of it, and if matter consequently did not possess them?

Pol. Certe consuevit dicere Aristoteles cum suis potius formas educi de potentia materiae, quam in illam induci; emergere potius ex ipsa, quam in ipsam ingeri: but I would say that Aristotle has preferred to call act the development of the form rather than its implication.

Dix. And I say that being expressed, sensible, and unfolded, is not the principal concept of actuality, but is a consequent and effect of that; in the same way as the principal being of wood, and the concept of its actuality, does not consist in being a bed, but in its being of such substance and consistency that it can arrive at being a bed, a bench, a beam, an idol, or whatever other thing that is constructed of wood. I prescind from this that it is according to a higher reason that natural things are made from natural matter, than artificial things are made from artificial matter: because art produces the forms from matter, either by substraction (as when it makes a statue out of stone) of by addition (as when, joining stone upon stone, earth, and wood, it constructs a house); but nature makes everything out of its matter by way of separation, birth, and effluxion, as the Pythagoreans understood, Anaxagoras and Democritus comprehended, and the sages of Babylonia confirmed. Moses also subscribed to this; Moses, who, in describing the generation of things ordered by the universal efficient cause employs this manner of speaking:

"The earth shall produce its animals, the waters shall produce the living animals"; as if he said: matter should produce them; because, according to him, the material principle of things is water, for which reason he says that the efficient intellect (called spirit by him) "walked on the waters"—that is, he gave to the waters a procreative virtue, and from them he drew out the natural species, all of which are called afterwards by him, in substance, waters. Whence, when speaking of the separation of inferior and superior bodies, he says that "the mind separated the waters from the waters," by means of which he suggests that the dry earth appeared. In this manner, therefore, all hold that things are made from matter by separation, and not by addition or reception. It is more appropriate to say, then, that matter contains the forms and implies them, than to think that it is empty of them and excludes them. That matter, then, which unfolds what it has enfolded must be called the divine and excellent progenitor, generator and mother of natural things; or, in substance, the entire nature. Is this not what you wish to affirm, Theophilus?

Theo. Certainly.

Dix. And it surprises me very much that our Peripatetics have not further developed the analogy of art. This, of the many matters that it knows and employs, always judges that that is better and more worthy which is less subject to corruption and more constant in its duration, and with which many more things can be made. Thus it holds gold to be more worthy than wood, stone, and iron, because it is less exposed to corruption, and because all that can be made of wood and stone can also be made of gold, and also many more, greater and better things, through its beauty, constancy, malleability, and nobility. And what must be said now of that matter of which man, gold, and all the other natural things are made? Must it not be considered more worthy than that of art, and must we not attribute to it greater actuality? Why, O Aristotle, do you not admit, that that which is the foundation and base of actuality, of that, I say, which is in act, and that which you declare to exist always—to exist in eternity—may be more in act than your forms, than your entelechies, that come and go, in such a manner, that if you should yet wish to find the permanence of this formal principle—

Pol. Because it is necessary that principles eternally remain.

Dix. —and not being able to recur to the fantastic ideas of Plato, which you hate so much, you will see yourself constrained and necessitated to say either that these specific forms have their permanent actuality in the hand of the efficient cause (and this you cannot say for the efficient is considered by you as that which draws out and extracts the forms from the potency of matter); or, they have their permanent actuality in the bosom of matter, and so you will be compelled to affirm; because all the forms that appear in the superficies of matter which you call individuals, and in act—as much those that were, as those that are, and will be—are originated things, and not the principle. And certainly I believe that the particular form is in the superficies of matter as the accident is in the superficies of the composed substance. For that reason lesser proportion of actuality must be accorded the expressed form with respect to matter; as, similarly, lesser proportion of actuality must be accorded to the accidental form with respect to the composite.

Theo. Truly, this is very poorly resolved by Aristotle, who says, together with all the ancient philosophers, that the principles must always remain permanent; and afterwards, when we seek in his doctrine where the natural form has its perpetual permanence (which fluctuates on the back of matter), we certainly do not find it in the fixed stars, because these particular forms that we see, do not descend from their height; nor in ideal signs, separated from matter, for we are certain that if these are not monstrosities, they are yet worse than monstrosities—I mean chimeras and vain fantasies. And then? The forms are in the bosom of matter. And then? She is the fountain of actuality. Do you wish that I should yet tell you more and make you see the great absurdity into which Aristotle has run? He says that matter is in potency. Ask him now when it will be in act. He will answer—and with him a great multitude—when it has the form. Add this question: What is that which possesses the new being? They will answer in spite of themselves: The composed, and not matter, because the latter is always the same, does not renew itself, and does not change. As in artificial things, when the statue is made out of wood, we do not say that there is added to the wood any new being, because it is no more wood now than it was before; but that which receives the being and actuality is that which is produced anew, the composition—that is, the statue. Now, then, how can you ascribe potency to that which will never be in act or will never have act? Therefore, matter is not in potency of being or that which can be, because it is always the same and immutable, and it is that about which, and in which, the mutation takes place, and not itself that which is changed. That which alters itself, enlarges itself, diminishes itself, changes its place, and is corrupted, is always, according to you Peripatetics, the composition, and never the matter. Why, then, do you say that matter is now in potency, now in act? Certainly, then, nobody should doubt that matter, either through receiving the forms, or through sending them forth from itself, does not receive more or less actuality, in relation to its essence or its substance. And, therefore, there is no ground through which it should be called in potency. This is more suitable to that which is in eternal rest and is even more the cause of that eternal rest; because if the form, in accordance with its fundamental and specific being, is of simple and invariable essence, not only logically in the concept and reason, but also physically in nature, it will have to be in the perpetual potency of matter—which is a potency indistinct from act, as I have explained in many ways, when I have discussed potency many times.

Pol. I beg you, say something of the *appetitus* of matter, to the end that we can find some resolution of a certain controversy between Gervasius and myself.

Gerv. Please do this, Theophilus; for this man has wearied my head with the analogy between matter and woman; he says that the woman does not content herself any less with men than the matter does with forms—and so forth, running on and on.

Theo. Since the matter does not receive anything from the form, why do you want it to desire it? If, as we have said, she sends the forms forth from her bosom and, consequently, has them in herself, why do you wish to say that she desires them? It does not desire those forms which daily change on her back, because every well-ordered thing desires that from which it receives perfection. What can a corruptible thing give to an eternal thing—an imperfect thing, as is the form of sensible things which is always in movement—to something so perfect that if it is well contemplated is a divine being in things—as perhaps David of Dinant, who was badly understood by those who reported his opinions, wished to say. It does not desire it to be preserved by it, because the corruptible thing does not preserve the eternal thing; moreover, it is evident that the matter preserves the form; whence such form should rather desire the matter in order to perpetuate itself; because when separated from it, it loses its being, and not matter which has all that which it had before the form was found, and which can also have others. Moreover, when the cause of corruption is given, it is not said that the form flees matter, or that it leaves matter, but rather that matter throws off that form, in order to take on another. Moreover, we haven't more cause to say that the matter desires the form than on the contrary to say that it hates it (I speak of those forms that are generated and corrupted, because the fountain of forms which is in itself it cannot desire, for the reason that nothing desires what it itself possesses); because through such reasoning, through which it can be said to desire what it sometimes receives or produces, it can be said, similarly, that when it throws it off and away, it hates it; nay, it detests more potently than it desires, for it eternally throws off that individual form which it retains for such a short time. If then you will remember this: that as many forms as it takes on, so many it also throws off, you must equally allow me to say that it has in itself a loathing—as I will allow you to say that it has in itself [a] desire.

Gerv. Here, then, the castles of Polyhymnius lie; together with those of others.

Pol. Parcius ista viris . . . .

Dix. We have learned enough for today. See you tomorrow.

Theo. Goodbye.

END OF THE FOURTH DIALOGUE

FIFTH DIALOGUE

Theo. The universe is, then, one, infinite, immobile. One, I say, is the absolute possibility, one the act, one the form or soul, one the matter or body, one the thing, one the being, one the greatest and the best—which must not be capable of being comprehended and, therefore, is without end and without limit—and in so far infinite and indeterminate—and consequently immobile. This does not move itself locally, because it has nothing outside of itself to which to transport itself—since it is itself all. It does not generate itself because there is no other being which it might desire or expect—(it) being that which has all the being. It is incorruptible because there is no other thing into which it could change itself—(it) being that which is everything. It cannot diminish or grow—(it) being infinite, the infinite being that which has no proportional parts, being that to which nothing can be added, and being that from which nothing can be subtracted. It is not changeable into any other disposition because it has no external through which it is passive and through which it can be affected. Besides which, through comprehending in its being all contrariety in unity and fitness, and not having any inclination to other and new being, or yet to any other mode of being, it cannot be the subject of change according to some quality, nor can it have any contrary or different thing which could alter it because in it everything coincides. It is not matter because it is not configurated, nor figurable; it is not limited, nor limitable; it is not form because it does not inform or figure anything else, being that which is all, greatest, one, universal. It is not measurable, nor does it measure. It does not include itself because it is not greater than itself; it is not included by itself because it is not less than itself. It is not equal to itself because it is not other and other, but one and the same. Being one and the same, it has not being and another being; and because it has not being and another being, it has not parts and again parts; and having no parts, it is not composed. It is a term in such a way that it is not a term; it is form in such a way that it is not form; it is matter in such a way that it is not matter; it is soul in such a way that it is not soul—because it is all indifferently and, in short, is one; the universe is one.

In it, most certainly, the height is not greater than the length and the depth; whence, through a certain similitude it is called, but is not, a sphere. In the sphere, the length, height, and depth are the same because they have the same end; but in the universe, length, height, and depth, are the same thing because in the same manner they have no term and are infinite. If they have no halves, quadrants, and other types of measure, if there is no measure, there is no proportional part, nor is there absolutely any part that differs from the whole. Because if you wish to speak of parts of the infinite, it is necessary to call it infinite; if it is infinite, it coincides in one being with the whole; therefore, the universe is one, infinite, and indivisible. And if the infinite does not find in itself difference, as part from whole, and other and other, the infinite certainly is one. Under the comprehension of the infinite there is no greater part and no lesser part because a greater part does not conform more to the proportion of the infinite than any other smaller part; therefore, in its infinite duration, the hour does not differ from the day, the day from the year, the year from the century, the century from the moment, because the moment and the hours are not more than the centuries, and those have not less proportion than those to eternity. Similarly, in the immensity, the foot is not different from the furlong, the furlong from the mile, because the mile does not more conform to the proportion of immensity than does the foot. Therefore, infinite hours are not more than infinite moments, or infinite feet than infinite miles. Thou canst not more nearly approach to a proportion, likeness, union, and identity with the infinite by being a man than by being an ant; not more nearly by being a star than by being a man; because you cannot more nearly approach that being by being sun, moon, than by being man, or an ant; for in the infinite these things are indifferent. And what I say of these, I mean to imply of all the other things of particular subsistence.

Whence, if all these particular things are not other and other in the infinite, are not different, are not species, they have, consequently, no number; therefore, the universe is one immobile thing. This because it comprises all, and is not affected by one and another being, and does not bear with itself nor in itself any mutation, is, consequently, all that [it] can be; and in it, as I have said the other day, the act is not different from the potency. If the potency is not different from the act, it is necessary that in it the point, the line, the surfaces, and the body are not different; for then, that line is surface, as the line, in moving itself, can become surface; then, that surface is moved and becomes a body, since the surface can be moved, and with its movement can become a body. It is necessary, then, that in the infinite, the point does not differ from the body because the point, running away from being a point, becomes a line; running away from being a line, it becomes a surface; running away from being a surface, it becomes a body; the point then, since it is in potentiality a body, does not differ from being a body, where the potency and the act are one and the same thing.

Therefore, the indivisible is not different from the divisible, the simplest from the infinite, the center from the circumference. Because then the infinite is all that it can be, it is immobile; because in it everything is indifferent, is one; and because it has all the greatness and perfection that can be had altogether, it is the maximum and the greatest immensity. If the point does not differ from the body, the center from the circumference, the finite from the infinite, the maximum from the minimum, surely we can affirm that the universe is all center, or that the center of the universe is everywhere, and that the circumference is not in any part, although it is different from the center; or that the circumference is throughout all, but the center is not to be found inasmuch as it is different from that. Here, then, as it is not impossible but necessary that the best, the greatest, the incomprehensible, is all, is throughout all, is in all, because as simple and indivisible, it can be all, it can be throughout all, it can be in all. And so it has not been vainly said that Jove fills all things, inhabits all the parts of the universe, is the center of all that has being—one in all, and through which one is all. Which, being all the things, and comprising in itself all being, brings it about that everything exists in everything.

But you would ask me: Why then do things change? Why does particular matter force itself to other forms? I answer you—there is no mutation that seeks another being, but rather another mode of being. And this is the difference between the universe and the things of the universe—because that comprises all the being and all the modes of being; and of the latter, each one has all the being, but not all the modes of being. And it cannot actually have all the circumstances and accidents because many forms are incompossible in the same subject, either through being contraries or through pertaining to different species; as there cannot be the same individual *suppositum* under the accidents of horse and man, under the dimensions of a plant and an animal. Whence, that [infinite] comprises all being totally, because outside and beyond the infinite being, there exists nothing that is, because it has no outside and no beyond; of these [things], each one comprises all the being, but not totally, because beyond each there are infinite others. Therefore, it is to be understood that all is in all, but not totally in all the modes in each one. Therefore, understand that everything is one, but not in the same mode.

Therefore, it is not an error to call entity, substance, and essence, one being, which, as infinite and undertermined, as much according to substance, as to duration, as to greatness, as to power, has not the function of principle, nor of the originated because, everything concurring in unity and identity—I say the same being—comes to have an absolute essence and not a relative one. In the immobile infinite, which is substance, which is being, there is found multitude and number which, through being a mode and multiplicity of being which denominates thing by thing, does not have the effect that being is more than one, but that it is [made] multiform, manifold, and multiplied. Therefore, profoundly considering this with the natural philosophers, and leaving the logicians in their fantasies, we find that all that makes difference and number is pure accident, pure figure, and pure complexion. Every production, of whatever kind that it may be, is an alteration, with the substance always remaining the same, because this is one—one divine and immortal being. Pythagoras, who does not fear death but awaits change, understood this. All philosophers, commonly called physicists, have understood this; for they say that nothing is generated or corrupted with regard to substance—unless we want to understand alteration in this way. Solomon, who says "that there is nothing new under the sun, but that which is, always was," has understood this. You have, therefore, this fact: that all things are in the universe, and the universe is in all things—we in that, that in us; and, therefore, all things concur in a perfect unity. You see by this, then, that we ought not to torment our spirit, for there is no thing by which we ought to become vexed. Because this unity is one and stable, and always remains; this one is eternal; every face, every figure, every other thing is vanity, and is like nothing; nay, all that which is outside this one is nothing. Those philosophers who have found this unity have found their friend, wisdom. Wisdom, truth, and unity are throughout, one and the same. All have known to say that truth, one, and being are the same thing, but not all of them have understood it, because some have followed the manner of speaking but have not comprehended the manner of meaning of the true wise men. Aristotle, among others, who did not find unity, did not find being, and did not find the truth, because he did nor recognize being as one; and though he has been free to take the significance of being as common to substance and accident, besides distinguishing his categories according to so many genera, and species according to so many differences, he has not been very clever in the truth, through not deeply considering the knowledge of this unity and the indifference of constant nature and being; and as a right shallow sophist, he perverts the meanings of the ancients and opposes the truth, with malignant explanations and fickle persuasions—not so much perhaps through the imbecility of intellect as through the force of jealousy and ambition.

Dix. Thus, this world, this being, this truth, this universe, this infinite, this immensity, is wholly in all its parts, and consequently is the (ubique) everywhere itself. Thence, that which is in the universe is through all in relation to the universe, or in relation to the other particular bodies, according to the mode of its capacity—because it is above, below, innermost, right, left, and according to all local differences; because in the whole infinite there are all these differences and no one of them. Everything that we find in the universe—because it has that in itself which is all throughout all—comprehends in its mode the whole world soul, although not totally, as we have formerly pointed out; the world soul is whole in whatever part of that [universe of things]. Therefore, as the activity is one, and constitutes one being—wherever it is—so one must not believe that there is in the world a plurality of substances, and a plurality of that which is truly being. Furthermore, I know that you have known as a fact that each one of all these innumerable worlds that we see in the universe is not in the universe—so much as in a containing place, and as in an interval and space—as much as they are in one comprehending, preserving, moving, efficient cause; therefore, the whole comes to be comprehended by each one of these worlds, as the whole soul is comprehended by each part of the same. Therefore, although a particular world moves itself toward and about the other, as the earth to and about the sun—in relation to the universe—nothing is moved toward or about that, but in that.

Further, you wish to say, that as the soul, even according to the usual opinion, is in all the great vastness to which it gives being, and is in itself indivisible, and consequently the same throughout all, and in whatsoever part entirely, so also the essence of the universe is one in the infinite, and in whatsoever thing taken as a member of that; so that it in fact, the whole and every part of that, comes to be one according to substance; therefore, Parmenides has not unsuitably called this one, infinite, and immobile: be it as you wish with his intention—which is uncertain because it has been related by a not too faithful historian.

You say that everything we see of difference in bodies, in relation to formations, complexions, figures, colors, and other properties or common qualities, is nothing else than a diversity of appearance of the same substance; a transitory, mobile, corruptible appearance of an immobile, stable and eternal being; in which all forms, figures, and members are but as indistinct and as conglomerated, not otherwise than it is in the seed, wherein the arm is not distinct from the hand, the bust from the head, the nerve from the bone. That distinction and separation does not come to produce new and other substance, but puts into activity and completion certain qualities, differences, accidents, and orders of that one substance. And that which is said concerning the seed, in relation to the members of animals, similarly can be said of food, as regards its being chyle, blood, phlegm, meat, seed, and similarly of the other things that precede its being food or any other thing; and similarly of all things; and thus, rising from the lowest grade of nature to the highest level of that, from the physical universe, known by philosophers, to the height of the archetypal universe, as believed by the theologians—if you please; until you arrive at an original and universal substance of the whole, which is called being, the foundation of all diverse forms and species; as in the art of carpentry there is a substance of wood, subject to all measures and figures, which are not the wood, but of wood, in wood, and about wood. Therefore, all that which makes for diversity of genus, species, difference, properties—all that which consists in generation, corruption, alteration, and change—is not being or existence, but a condition and circumstance of being and existence, which is one, infinite, immobile, subject, matter, life, soul, true, and good.

It is also your wish, then, that since the being is indivisible and simple, because it is infinite, and all act in all, and all in every part (in such a way that we say a part in the infinite, not part of the infinite), we cannot think in any way that the earth is a part of being, the sun a part of substance—as that is indivisible; but it is legitimate to say substance of the part, or better still, substance in the part; thus, as it is not legitimate to say that a part of the soul is in the arm, and a part of the soul is in the head, it is legitimate to say that the soul is in the part, which is head, and the substance is in the part or of the part, which is arm. Therefore, to exist, as portion, as part, as member, as whole, as equal, as greater, as smaller, as this, as that, of this, of that, as coordinated, as different (and other reasons), does not signify an absolute and therefore, cannot be referred to as substance, one, and being, but exists through substance, in the one and about being, as modes, reasons, and forms; as it is usually said that about a substance there is quantity, quality, relation, action, passion, and other circumstances of genus: thus, the one highest being, in which the act is indifferent from the potency—which can be all absolutely—and is all that it can be, is complicatively one, immense, infinite, and comprises all being; and is explicatively in these sensible bodies and in the distinct potency and act, which we see there in these things. Therefore, you wish to say, that that which is generated and generates (either be it an equivocal or a univocal agent, as those who philosophize commonly say) and that of which the generation is made are always of the same substance. And thus, it will not sound hard to your ears to hear the statement of Heraclitus, who said that all things are one, which through its mutability has in itself all things; and since all the forms are in it, consequently all definitions are conformable to it, and all contradictory propositions are true. And that which makes for multiplicity in things is not being, is not the thing, but that which appears, that which is represented to the senses, and that which is in the surfaces of the thing.

Theo. Correct. I want, however, that beyond this you grasp more points of this very important science, and of the very solid foundation of the truths and secrets of nature. First, then, I want you to note that there is one and the same scale, through which nature descends to the production of things, and the intellect ascends to the cognition of them, and that one and the other proceeds from unity to unity, passing through the multiplicity of media. I submit that with their mode of philosophy the Peripatetics and many Platonists, have the multitude of things as a means, preceded by the purest act at one extreme, and by the pure potency at the other; as others, through a certain metaphor, wish to allow the darkness and the light to work together towards the constitution of innumerable grades of forms, effigies, figures, and colors; after those who consider two principles and two leaders occur other enemies and adversaries of polyarchy, and make these two concur into one, which is similarly abyss and darkness, clarity and light, profound and impenetrable obscurity, high and inaccessible light.

Second, consider that the intellect, wishing to free itself and loosen itself from the imagination to which it is joined, besides recurring to mathematical and imaginable figures in order that it may either through those or through similarities to that understand the being and substance of things, also comes to refer the multitude and diversity of species to one and the same root; as Pythagoras, who posited the numbers as the specific principles of things, understood unity as the foundation and substance of all; [as] Plato and others, who placed the stable species in the figures, understood the point as the trunk and root of all, as substance and universal genus; and perhaps the surfaces and figures are those that Plato at the end understood by his "Great," and the point and atom is that which he understood by his "Small"—double specific principles of things which are then reduced to one, as every divisible is reduced to the indivisible. Those, then, who say that the substantial principle is one take the substances as numbers; the others who understand the substantial principle as the point take the substances of things as figures; and all agree to set an indivisible principle. But better still and purer than Plato's, is the method of Pythagoras, because unity is the cause and reason of indivisibility and punctuality, and is a principle more absolute and more conformable to the universal being.

Gerv. How is it that Plato, who came after Pythagoras, did not do something similar or better than Pythagoras?

Theo. Because he wished, though speaking worse and in a less fitting and less appropriate way, to be judged a master rather than, by speaking on a better doctrine and in the best way, to be judged a disciple. I want to say that the end of his philosophy was more his own fame than truth; yet I cannot doubt that he knew very well that his doctrine was more suitable to things corporeal and things corporeally considered, while the other doctrine was as good and as suitable for these—as yet to all the others—which reason, imagination, intellect, one and another nature, knew how to produce. Everyone will admit that is was no secret to Plato that unity and number are indispensable to examine and understand the points and the forms; but, on the other hand, the forms and points are not indispensable in order to reach an understanding of unity and number, as dimensional and corporeal substance depends on the incorporeal and the indivisible; moreover, this is absolute from that because the understanding of numbers is found without that of measure, but that cannot be absolute from this because the understanding of measure cannot be found without that of number; therefore, the arithmetical analogy and proportion is more conformable than the geometrical to lead us through the medium of the multitude to the contemplation and apprehension of that indivisible principle. Because it is the unique and radical substance of all things, it is impossible that it should have a certain and determined name and such a term that signifies positively rather than privatively; and, therefore, it has been called by some "point," by others "unity," by others "infinity," and by others variously according to other concepts similar to these.

Add to what has been said that when the intellect wishes to comprehend the essence of a thing, it simplifies as much as possible; I wish to say, that is, that it retires from composition and multiplicity, by throwing off the corruptible accidents, the dimensions, the signs, and the figures, from that which lies under these things. Thus, we do not understand the long and prolix oration except by contraction to a simple intention. The intellect in this openly shows that the substance of things consists in unity, which it seeks either in similitude or in truth. Believe that that would be the most consummate and the most perfect geometrician who could contract to one intention only all the intentions dispersed in the principles of Euclid; likewise, that would be the most perfect logician who had all the intentions contracted into one. Herein is given the degree of the intelligences: the inferior cannot understand many things except through many species, analogies, and forms; the superior ones understand better with less; the highest understand perfectly with the least. The first intelligence comprehends everything in one perfect idea; the divine mind and the absolute unity, without any species at all, is itself, at once and the same, that which understands and that which is understood. So that we, in rising to the perfect cognition, proceed by simplifying the manifold; just as, in descending to the production of things, unity proceeds by unfolding itself. The descent is from the one being to infinite individuals and innumerable species; the ascent is from these to that.

To conclude, therefore, this second consideration, I say that when we aspire toward and we strive for the first principle and substance of things, we make progress towards indivisibility; and we never believe to have attained that first being and universal substance so long as we have not arrived at that one indivisible in which everything is comprehended; and so likewise we believe that we cannot understand more of substance and essence than we can understand of indivisibility. Hence, the Peripatetics and the Platonists reduce many species to one indivisible reason; they comprehend innumerable species under determinate genera—as Archytas, who first set forth ten—determinate genera leading to one being, one thing, which thing and being is understood by them as a name and a term and a logical intention, and, finally, as a vanity. Therefore, afterwards, when they treat of physics, they do not know a principle of reality and being of all that which is as an intention and common name to all that which is said and comprehended; and this, surely, has happened because of the weakness of the intellect.

Thirdly, you ought to know that since substance and existence are distinct and absolute from quantity, and measure and number are consequently not substance, but about substance, not being, but things of being, it follows that we ought necessarily to say that substance is essentially without number and without measure and, therefore, one and indivisible in all particular things; these latter things have their particularity from number, that is, from things that are about substance. Wherefore, he who apprehends Polyhymnius, as Polyhymnius, does not apprehend particular substance, but substance in the particular and in the differences, which are about that; the substance, through the medium of the latter, places this man under a species in number and multitude;

herein, just as distinct accidents cause multiplication of these which are called individual examples of humanity, so distinct accidents of animals cause multiplicity of these species of animality. Likewise, distinct accidents of life cause multiplication of this living and animated organism. It is not otherwise that distinct corporeal accidents cause multiplication of corporeality. Similarly, certain accidents of substance cause multiplication of substance. In such a manner, certain accidents of existence cause multiplication of entity, of truth, of unity, of being, of the true, of the one.

Fourthly, take notice of the signs and the verifications, by means of which we wish to conclude that the contraries coincide in one; whence it is not made difficult to infer that all things are one, as every number, just so even or odd, just so finite or infinite, is reduced to unity—which repeated in the finite series posits number, and repeated in the infinite series negates number. You shall take the signs from mathematics, and the verifications from the other moral and speculative faculties. Hence, as regards the signs, tell me: What thing is more unlike the straight line than the circle? What thing is more contrary to the straight line than the curve? Yet they coincide in the principle and in the smallest part; since, as Cusanus, the discoverer of the most beautiful secrets of geometry, has divinely observed—What difference will you find between the smallest arc and the smallest cord? Further, in the greatest; what difference will you find between the infinite circle and the straight line? Do you not see that as the circle grows larger it more and more approaches with its activity the straight line? Who is so blind that he does not see that inasmuch as the arc BB, because it is greater than the arc AA, and the arc CC greater than arc BB, and the arc DD greater than the other three, tend to be parts of larger circles; and thus approach more and more to the

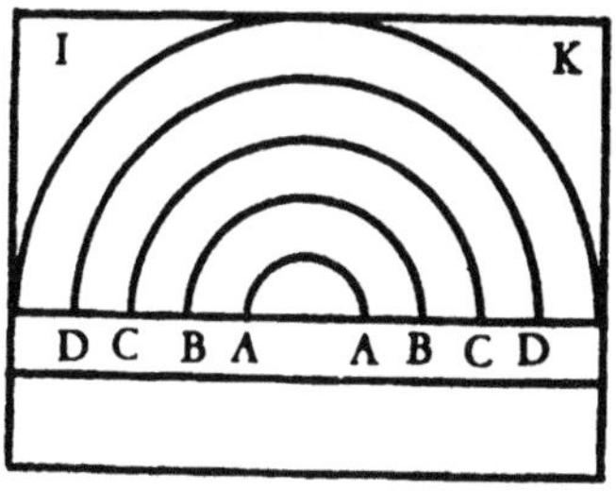

Fig. 10

straightness of the infinite line of the infinite circle, signified by IK? Here one must most surely believe and say that just as that line which is greater according to the concept of greater size is also straighter, so, similarly, the greatest of all must be in the superlative the straightest of all; so much so, that in the end, the infinite straight line becomes the infinite circle. Here, therefore, not only do the maximum and the minimum coincide in one being, as we have demonstrated so many times, but also the contraries coincide as one and indifferent in the maximum and the minimum. Besides, if you want to compare the finite species to the triangle, since all finite things are understood to participate in finiteness and termination, by a certain analogy, through the first finite and the first limited (as all the analogous predicates in all the genera take their grade and order from the greatest and first of that genus), and inasmuch as the triangle is the basic figure which cannot be resolved into any other kind of simpler figure (as, on the contrary, the quadrangle can be resolved into triangles) and is, therefore, the first foundation of every limited and configurated thing, you will find that the triangle, as it cannot be resolved into any other figure, likewise cannot proceed in triangles in which the three angles are greater or lesser, although they may be various and diverse—of various and diverse figures—as to greater or lesser magnitude—even to the greatest and the least. Therefore, if you take an infinite triangle—I do not mean really or absolutely because the infinite has no figure, but an infinite triangle, by supposition, and with an angle that is appropriate for that which we wish to demonstrate—it will have no greater angle than the smallest finite triangle; no

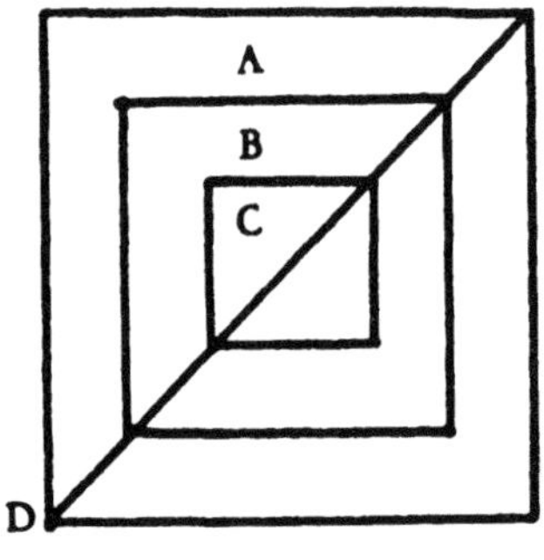

Fig. 11

greater angle than any of the intermediary ones or the greatest one.

Yet leaving aside the comparison of figures and figures—I mean of triangles and triangles, and taking angles and angles, whether great and small—they are all seen to be equal, as appears in this square. The latter, is divided through the diameter into so many triangles: whence it is seen that, not only are the right angles of the three squares, A, B, C equal, but also all the acute angles, which are the result of the division of the said diameter—which constitutes twice as many triangles, all of equal angles. Thence, through very intelligible analogy, one sees how the one infinite substance can be all in all things, albeit in some finitely, in others infinitely—in this with less, in that with greater measure.

Add to this (to see further that the contraries coincide in this one and infinite) that the acute angle and the obtuse angle are two contraries, which, don't you see, are produced from one and the same principle—that is, from an inclination, which the perpendicular line M makes, which perpendicular is joined to the horizontal line BD, at the point C?

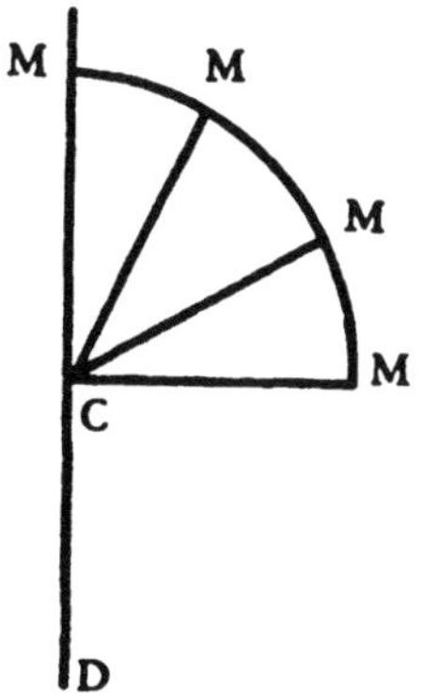

Fig. 12

This, at that point, with a simple inclination toward the point D, after it had produced the right angle and the right angle indifferently, produces a greater difference of acute and obtuse angles as it advances toward point C, to which, having been joined and united, it produces the indifference of the acute and the obtuse, similarly annulling the one and the other because they are one in the potency of the same line. That, since it has been able to be united to, and to make itself indifferent with the line BD, can thus disunite itself and make itself different from that, evolving the most contrary angles from one and the same indivisible principle—angles which range from the maximum acute and the maximum obtuse to the minimum acute and the minimum obtuse—and thence to the indifference of the right angle and to that coincidence which is constituted by the contact of the perpendicular and the horizontal line.

To the means of verification. First of all, concerning the prime active qualities of corporeal nature, who does not know that the principle of heat is indivisible and, therefore, separated from every kind of heat, because the principle must not be something of the originated? If this is so, who can object to the affirmation that the principle is not warm, nor cold, but an identity of warmth and cold? Whence does it happen, then, that one contrary is the principle of the other—and, therefore, that the transformations are circular—unless there exists a substratum, a principle, a term, and a continuation and a coincidence of one and the other? Are not the minimum warmth and the minimum cold the same throughout? Isn't the principle of movement toward the cold to be found in the limit of the maximum warmth? It is obvious, therefore, that not only do the two maximums (maxima) sometimes coincide in resistance, and the two minima (coincide) in agreement, but also the maximum and the minimum through the vicissitude of transmutation; therefore, it is not without good cause that the doctors are apprehensive of the perfect state of health; in the highest level of felicity the provident are cautious. Who does not see that the principle of generation and corruption is one? Is not the end of corruption the principle of generation? Do we not similarly say that taken, this placed; that was, this is? Certainly, if we consider well, we see that corruption is not other than a generation, and the generation is not other than a corruption; the love is a kind of hate; (finally) the hate is a kind of love; the hate of the contrary is the love of the convenient; the love of this is the hate of that. In substance and root, therefore, love and hate, friendship and conflict, is one and the same thing. From whence does the doctor seek the antidote more fittingly than from venom? What gives better treacles than the viper? In the worst poisons [are] the best medicines. Is not the same potency common to two contrary objects? Now, where do you believe this comes from, if not from that; that as one is the principle of being, so one is the principle of conceiving one and the other object; and that the contraries are about one substratum, as they are apprehended by one and the same sense? Not to speak of that—that the circular rests on the level, the concave tarries and lies on the convex, the irascible lives together with the patient, the humble is pleasing to the most arrogant, the liberal to the avaricious.

In conclusion, he who wishes to know the greatest secrets of nature should regard and contemplate the minimum and maximum of contraries and opposites. It is profound magic to know how to draw out the contrary after having found the point of union. The poor Aristotle directed his thought to this, establishing privation, to which is joined a certain disposition, as progenitor, parent, and mother of form; but he could not attain the end. He has not been able to arrive at this point because he stopped at the genus of opposition; he remained entangled in such a manner that in not descending to species from contrariety, he did not arrive, nor did he fix his eyes upon the goal; therefore, he has, with this one assertion—that contraries cannot actually harmonize in the same subject—missed the entire way.

Pol. You have discoursed sublimely, rarely, and extraordinarily of the all, of the maximum, of the being, of the principle, and of the one. But I would like to see you show forth the difference of unity; because I find it written: "It is not good to be alone." Besides, I feel great anxiety because there, in my purse and money bag, but one penny is lodged.

Theo. That unity is all which is not unfolded, not under distribution and distinction of number, and does not exist in such singularity—as you would like to understand it—but is complicative and comprising.

Pol. An example? Because to tell the truth, I hear, but I do not understand.

Theo. Just as the decade is a unity, but is embracing, so the hundred is not less a unity, but is more embracing; the thousand is a unity no less than the others, but is still more embracing. This that I propose to you in arithmetic, you ought to understand, in a higher and simpler sense, of all things. The highest good, the highest object of desire, the highest perfection, the highest beatitude, consists in the unity that embraces all. We take delight in colors but not in one express color, whatever that may be, but the greatest delight is in such a one that embraces all colors. We delight in sound, not in a particular one, but in an embracing one which results from the harmony of all. We delight in a sensible, but greatest delight is in that which comprehends all the sensibles; we delight in a knowable that comprises everything knowable; we delight in the apprehensible that embraces all that which can be comprehended; we delight in a being which embraces all, but greatest delight is in that one which is the all itself. As you, Polyhymnius, would be more delighted in the unity of a gem so precious that it would be more valuable than all the gold in the world than in the multitude of thousands and thousands of such pennies as the one which you have in your pocketbook.

Pol. Excellent.

Gerv. I am now too a learned person; because just as he who does not understand the one does not understand anything, so he who understands the one truly understands all; and he who advances more to the knowledge of the one, more nearly approaches the knowledge of all.

Dix. And so go I—if I have understood well—being tremendously enriched by the thoughts of Theophilus, faithful reporter of the Nolan philosophy.

Theo. Praised be the Gods, and extolled by all the living be the infinite, the simplest, the most unified, the highest, and the most absolute cause, principle, and the one.

End of the Fifth Dialogue

NOTES

INTRODUCTION

1. For Aristotle, it is the knowledge of the nature that is being sought. The study of operations is the means to achieve this knowledge. Aristotle, *De anima*, i. l. 402ab15–25; *Metaphysics*, ii. 2. 996b15–19. All references to Aristotle are from Aristotle, *Works*, Oxford Edition (1910–1925), translated and edited by W. D. Ross.

2. Bartholmess, *Jordano Bruno* (1846–1847), 2 vols.

3. *Carrière, Die philosophiche Weltanschauung der Reformationszeit, in ihren Beziehungen zur Gegenwart* (1847; 2d ed., 1887).

4. Berti, *Vita di Giordano Bruno da Nola* (1868) and *Documenti intorno a Giordano Bruno* (1880).

5. Clemens, *Giordano Bruno und Nicolaus von Cusa* (1847).

6. Brunnhofer, *Giordano Brunos Weltanschauung und Verhangniss aus den Quellen dargestellt* (1882).

7. Wagner, *Opere di Giordano Bruno*, Ora per la prima volta raccolte e pubblicate da Adolfo Wagner (1830), 2 vols.

8. Tocco, *Le opere latine di Giordano Bruno esposte e confrontate con le italiane* (1889) and "Le fonti piu recenti dell filosofia del Bruno," *Rendiconti della Reale Accademia dei Lincei*, I, Series 5 (1892), 503–581, 585–622.

9. Gentile, *Il pensiero italiano del rinascimento* (1940), pp. 259–330.

10. Mondolfo, "La filosofia di Giordano Bruno e la interpretazione di Felice Tocco," *La cultura filosofica*, V (1911), 450–482.

11. Troilo, *La filosofia di Giordano Bruno* (1907).

12. Olschki, *Geschichte der neusprachlichen Wissenschaften Literatur* (1927), III, 1–67; *Giordano Bruno* (1927); "Giordano Bruno," *Deutsche Vierteljahrschrift fur Literaturwissenschaftlichen und Geistesgeschichte*, II (1924) 1–78.

13. Spampanato, *Vita di Giordano Bruno* (1921).

14. Salvestrini, *Bibliografia della opere di Giordano Bruno* (1926).

15. Namer, *Les aspects de Dieu dans la philosophie de Giordano Bruno* (1926).

16. Sarauw, *Der Einflus Plotins auf Giordano Brunos Degli Eroici Furori* (1916).

17. De Ruggiero, *Giordano Bruno* (1913).

18. Corsano, *Il pensiero di Giordano Bruno* (1940).

19. Guzzo, *Giordano Bruno, De la causa, principio e uno* (1933).

20. Cassirer, *Individuum und Kosmos in der Philosophie der Philosophie der Renaissance* (1927).

21. McIntyre, *Giordano Bruno* (1903).

22. Frith, *Life of Bruno* (1887).

23. Jordani Bruni Nolani, *Opera latine conscripta*, edited by F. Fiorentino, F. Tocco, H. Vitelli, V. Imbriani, and C. M. Tallarigo (1879–1891), 3 vols., 8 parts.

24. Bartholmess, *Jordano Bruno*, II, 251 ff.

25. *Ibid.*, p. 253.

26. John Toland, *Collection of Several Pieces of Mr. John Toland;* with some memories of his life and writings (1726), Vol. I. This work contains a translation of the introductory epistle to the "De l'infinito."

27. Cf. Carrière, *Die philosophiche Weltanschauung der Reformationszeit*, p. 487.

28. Schelling, *Bruno, oder uber das gottliche und naturliche Princip der Dinge* (1802).

29. Hegel, *Lectures on the History of Philosophy*, translated by E. S. Haldane and F. H. Simpson (1892–1895), III, 119–137.

30. Frith, *Life of Bruno*, pp. viii-ix.

31. Bartholmess, *Jordano Bruno*, II, 251–316; cf. *ibid.*, p. vi.

32. This contribution has also been recognized by Mrs. Frith; see *Life of Bruno*, p. ix.

33. Brunnhofer, *Giordano Brunos Weltanschauung*, pp. 151–154; cf. *ibid.*, p. 169.

34. McIntyre, *Giordano Bruno*, p. 354; cf. *ibid.*, pp. 333, 337.

35. Brunnhofer, *Giordano Brunos Weltanschauung*, pp. 135–309. Brunnhofer insists that Goethe, Galileo, and Spinoza owed a great deal of their teachings to Bruno; in fact, Brunnhofer holds that Spinoza's philosophy would have been impossible without Bruno.

36. Tocco, *Le opere latine di Giordano Bruno*, pp. 327–416.

37. McIntyre, *Giordano Bruno*, p. viii.

38. Olschki, *Geschichte der neusprachlichen Wissenschaften Literatur*, III, 55–56; cf. Olschki, *Giordano Bruno* (1927), p. 51.

39. Frith, *Life of Bruno*, pp. 16–17.

40. Tocco, *Le opere latine di Giordano Bruno*, pp. 331–332; cf. *ibid.*, p. 357.

41. *Ibid.*, p. 334.

42. *Ibid.*, pp. 335–336: "Per lo che anche nel Sigillus, come nel De Umbris, it fondo della dottrina, le immagini che l'adombrano, la terminologia tecnica, lutto è schiettamente Neoplatonico."

43. *Ibid.*, p. 337.

44. *Ibid.*, p. 342: "Non possendo ricorrere alle fantastiche idee di Platone."

45. *Ibid.*

46. Ibid., p. 347.

47. *Ibid.*, pp. 347–350.

48. *Ibid.*, pp. 352–354.

49. *Ibid.*, pp. 356–357.

50. *Ibid.*, pp. 357–361.

51. Gentile, *Il pensiero italiano del rinascimento*, pp. 311–330.

52. *Ibid.*, pp. 259 ff.

53. Olschki, *Geschichte der neusprachlichen Wissenschaften Literatur*, III, 54–56. Cf. McIntyre, *Giordano Bruno*, p. viii: "Bruno's style, full as it is of obscurities, redundances, repetitions, lends itself to selection, but not easily to exposition."

54. Olschki, *Giordano Bruno* (1927), pp. 11–20. Cf. Mondolfo, "La filosofia di Giordano Bruno," pp. 450–482; especially pp. 458 ff., where the relationship

between "mens insita omnibus," and "mens super omnia," parallels the Gentile statements given above, though the point of view is directly opposed to Gentile's.

55. Tocco, *Le opere latine di Giordano Bruno*, pp. 211–326.

56. Cf. Chapter I below, p. 00.

57. Cf. St. Bonaventure, *In sent.*, i. 1, 3, 2, *ad resp.*; cf. Duns Scotus, *Op. Ox.*, i. dist. II, q. 2, n. 20.

58. "De immenso et innumerabilibus," *Opera latine conscripta*, I, Part I, 205.

59. *Ibid.*, p. 203.

60. Cf. Plato, *Phaedrus*, 6. 12.

61. Cf. Plato, *Republic*, Bks. vi–vii.

62. *Opere italiane*, edited by G. Gentile and B. Spampanato (1907–1909), I, 341. Cf. Plotinus, *Enneads*, translated by Stephen McKenna (1917–1930), III, 36–37 (Ennead iv. 3. 24).

63. *Opere italiane*, II, 341 ff.

64. *Ibid.* Cf. *ibid.*, p. 352.

65. *Ibid.*, p. 340: "How can our finite intellect pursue the infinite object? With the infinite power that it possesses . . . for our finite intellect pursues the infinite object because the human mind is eternal, and therein is its delight; and it has neither end nor measure in its felicity." Cf. Bartholmess, *Jordano Bruno*, II, 124.

66. *Opere italiane*, II, 349; cf. *ibid.*, p. 357; cf. "De immenso et innumerabilibus," *Opera latine conscripta*, I, Part II, 146.

67. *Opere italiane*, II, 357: "To be seen by God is to become one with God." Cf. *ibid.*, p. 349: "We have not to look for Divinity at a distance from us, for we have it with us, more truly intimate to us than we are to ourselves."

68. "De immenso et innumerabilibus," *Opera latine conscripta*, I, Part II, 1.

69. Aristotle, *Physics*, iii. 4. 203b15 ff.

70. *Ibid.*, 204a1–7.

71. *Ibid.*, 203a2–10.

72. *Ibid.*, 5. 204a20–30.

73. *Ibid.*, 204a32–35.

74. The Pythagoreans maintained a duality of principles, but added the notion that the limited (odd) the unlimited (even), and the one (unity) are not predicates of some other entity, like fire, air, water, etc., but are themselves the substance of things of which they are predicated.

75. Aristotle, *Physics*, iii. 4. 203a15–17.

76. Aristotle, *Metaphysics*, i. 6. 987b20–22.

77. *Ibid.*, 987b23–25.

78. *Ibid.*, 987b25–30.

79. *Ibid.*, 9. 992b, gives the detailed arguments against Plato. Here we are concerned with Plato only in relation to his position alongside the Pythagoreans.

80. Aristotle, *Physics*, iii. 5. 204a7–9; cf. *ibid.*, 204a20–33.

81. *Ibid.*, 4. 203a16–18.

82. *Ibid.*, 203a19–23.

83. *Ibid.*, 6. 207a1 ff.

84. *Ibid.*, 207a1 ff.

85. Cf. Mondolfo, *L'infinito nel pensiero dei greci* (1934); cf. Aliotta, "Il problema dell'infinito," *La cultura filosofica*, V (1911), 205-232.

86. Lucretius, *De rerum natura*, Bk. I, ll. 968-973, 977-979. All references are from H. A. J. Munro's translation (1919).

87. *Ibid.*, ll. 1052-1076.

88. Gilson, *The Spirit of Medieval Philosophy* (1940), p. 56.

89. *Ibid.*, p. 55. Cf. Aquinas, *Compendium theologiae*, chap. xx; *Comm. in libr. de causis*, Lectio 6; *De pot.*, 7. 2ad9; *Summa theologica*, 1, 3, 4, and 1, 4, 1, ad 3.

90. Gilson, *The Spirit of Medieval Philosophy*, pp. 56-57.

91. Cf. Harris, *Duns Scotus* (1927), II, 159 ff. Cf. Scotus, *De primo principio*, chap. iv, n. 21; chap. iv, n. 15; chap. iv, n. 23. Cf. Scotus, *Opus oxoniense*, i. dist. II, q. 2, nn. 27, 30, 31, cited by Harris, pp. 161 ff. Gilson, *The Spirit of Medieval Philosophy*, pp. 55-58.

92. Aristotle, *Physics*, iii. 6. 207a7-8.

93. Gilson, *The Spirit of Medieval Philosophy*, p. 55.

94. For the relationship between corporeal reality and incorporeal reality in the writings of Cusanus, see below, Chapter IV. This will be shown to be especially important because, for Bruno, the immaterial does not exist as immaterial, though he does not deny the existence of the incorporeal. Cf. Tocco, *Le opere latine di Giordano Bruno*, pp. 342 ff.

95. *Opere italiane*, I, 291.

I UNIVERSAL SOUL AND UNIVERSAL FORM

1. "De la causa" is divided into five dialogues, the first of which is an apology for the "La cena de le ceneri," which was written by Bruno in praise of the Copernican theory. Since it went far beyond Copernicus in its conclusions, it evoked sharp criticism from Bruno's opponents. These critics sought from Bruno some justification for his conclusions concerning the infinity of the universe and its innumerable worlds and for the claim that all these worlds were inhabited by living beings such as those that inhabit this planet. The "apology" takes the form of an argument "ad hominem," and leaves the philosophical answers to the remaining four dialogues. Hence the first dialogue bears no direct relationship to the material that follows, and it is for this reason that our investigation begins with the second dialogue of "De la causa." Each of the four persons who appears in the dialogues represents a school of thought. Bruno is played by Theophilus, who "serves to distinguish, to define, and to demonstrate the material"; Alexander Dixon is the representative of the Neoplatonic school, and it is he "whom the Nolan loves as his own eyes, and he who has made the planning of the work possible"; Gervasius is "not a philosopher by profession," but it is he "who assists our discussions with entertainment"; lastly, there is Polyhymnius, "who is a sacrilegious pedant, and one of the most rigid censors of philosophers." Our method will be to follow the basic context of the dialogues;

the goal will be to pursue the thread of thought which is woven by the individual speakers, without specific reference to the particular one in whose mouth the statements are placed. We shall, therefore, omit the material which is not essential to the evaluation and interpretation of the main doctrine. These omissions will be noted as they occur, so that the analysis does not lose contact with the text at any point.

2. *Opere italiane*, I, 168: "Or come intendete che le cose, che hanno causa e principio primo e prossimo, siano veramente conosciute, se, secondo la raggione della causa efficiente (la quale è una di quelle che concorreno alla real cognizione de le cose), sono occolte?"

3. *Ibid.*, p. 169: "Perchè dalla cognizione di tutte cose dependenti non possiamo inferire notizia del primo principio e causa, cheper modo men efficace che di vestigio, essendo che il tutto deriva dalla sua bontà o voluntà, la quale è principio della sua operazione, da cui procede l'universale effetto." The redundant "che" appears in the Lagarde and in the Wagner editions too; it must therefore be regarded as Bruno's, though the translation is made the more difficult by its presence. "Tanto che conoscere l'universo, e come conoscere nulla dello esser e sustanza del primo principio, perchè è come conoscere gli accidenti degli accidenti."

4. *Ibid.*, pp. 170–171.

5. Indeed, we shall see that for Bruno, Principle and Cause are inconceivable, not because they are above nature, but because they are confused with nature. Cf. Tocco, *Le opere latine di Giordano Bruno*, p. 347.

6. *Opere italiane*, I, 171: "Lasciando dunque, come voi dite, quella considerazione, per quanto è superiore ad ogni senso o intelletto, consideriamo del principio e causa, per quanto, in vestigio, o è natura istessa, o pur riluce ne l'ambito e grembo di quella."

7. *Ibid.*, p. 172: "Credo che vogliate che principio sia quello che intrinsecamente concorre alla constituzione della cosa e rimane nell'effeto . . . causa chiami quella che concorre alla produzione delle cose esteriormente, ed ha l'essere fuor de la composizione, come e l'efficiente e il fine, al quale e ordinata la cosa prodotta." Cf. "Summa terminorum metaphysicorum," *Opera latine conscripta*, I, Part IV, 17. Cf. Aristotle, *Metaphysics*, iv. 2. 1013a24–1013b3; *ibid.*, xii. 4. 1070b22–35; *Physics*, ii. 194b1–195a3.

8. *Opere italiane*, I, 175.

9. *Ibid.*, p. 175; cf. *ibid.*, p. 172.

10. *Ibid.*, p. 175; cf. *ibid.*, p. 172.

11. Bruno distinguishes three types of intellect at this point. Since he later shows, however, that the distinction is one of function, and not one of being, we are withholding comment on this point until the text again emphasizes it. Here, the final goal is to arrive at a definitive treatment of Cause and Principle; the world intellect is employed as a means to achieve this end, and the distinction between the different kinds of intellects would only lead to confusion if entered into in detail at this time.

12. *Opere italiane*, I, 171: "Rispondo che, quando diciamo Dio primo principio e prima causa, intendiamo una medsima cosa con diverse raggioni."

13. Cf. Tocco, *Le opere latine di Giordano Bruno*, p. 347. Tocco's entire argument is based upon the fact that "De la causa" mentions Neoplatonism only to depart sharply from it.

14. Plotinus, *Enneads*, III, 8 (Ennead iv. 3. 2).

15. *Ibid.*

16. *Ibid.*, p. 9.

17. *Ibid.*

18. *Ibid.*

19. *Ibid.*, p. 10 (Ennead iv. 3. 3).

20. *Ibid.*

21. *Ibid.*, p. 11.

22. *Ibid.*, p. 12 (Ennead iv. 3. 4).

23. *Ibid.*, V, III (Ennead iv. 4. 4).

24. *Ibid.*, p. 112.

25. The fact that Bruno leaves out, at this point, any mention of the material cause does not mean that he is not to employ all four of the Aristotelian causes. That it is left out in this context merely lends credence to the view stated above; namely, that here he is interested in treating only of the aspect of form, which will later be shown to be identical with the aspect of matter. *Opere italiane*, I, 171–173. Cf. "Summa terminorum metaphysicorum," *Opera latine conscripta*, I, Part IV, 17: "Causa per se est quadruplex famose: materia, forma, efficiens, finis, quibus Plato addidit locum et tempus." Cf. Aristotle, *Physics*, ii. 3.

26. *Opere italiane*, I, 173: "L'intelletto universale è l'intima, più reale e propia facultà e parte potenziale de l'anima del mondo."

27. *Ibid.*: "Quest e uno medesimo, che empie il tutto, illumina l'universo e indrizza la natura a produre le sue specie come si conviene." Cf. *ibid.*, pp. 250–251; "Spaccio de la bestia trionfante," *ibid.*, p. 13; "Summa terminorum Metaphysicorum," *Opera latine conscripta*, I, Part IV, 123.

Cf. McIntyre, *Giordano Bruno*, p. 157: "The intellectus is both internal and external to any particular thing: that is, it is not a part of any particular existence, is not exhausted by it; and, therefore, is so far external to it; on the other hand, it does not act upon matter from without, but from within; the efficient cause is at the same time an inward principle."

28. *Opere italiane*, I, 174: "Da noi si chiama artifice interno, perchè forma la materia e la figura da dentro"; that is, "We call this intellect the inner artificer because it shapes matter and configurates it from within." Again: "I call it [the universal intellect] extrinsic cause because as efficient it does not form a part of the things composed and the things produced; I call it intrinsic, in so far as it does not work around and outside of matter." Cf. "Summa terminorum metaphysicorum," *Opera latine conscripta*, I, Part IV, 73, 75; cf. "De monade, numero et figura," *ibid.*, Part II, pp. 238 ff. and pp. 215 ff.; *Opere italiane*, I, 296 and 300.

29. *Opere italiane*, I, 175: "Quanto, dico, più grande artefice è questo, il quale

non è attaccato ad una sola parte de la materia, ma opra continuamente tutto in tutto?"

30. *Ibid.*, pp. 175–176: "Mi par ch'abbiate a bastanza parlato della causa efficiente. Or vorei intendere che cosa è quella che volete sia la causa formale gionta all'efficiente: è forse la raggione ideale? Perchè ogni agente, che opra secondo la regola intellettuale, non procuraeffettuare, se non secondo qualche intenzione; e questa non è senza apprensione di qualche cosa; e questa non è altro che la forma de la cosa che è prodursi: e per tanto questo intelletto che ha facultà di produre tutte le specie, e cacciarle con si bella architettura della potenza della materia a l'atto, bisogna che le preabbia tutte, secondo certa raggion formale, senza la quale l'agente non potrebe procedere alla sua manifattura; come al statuario non e possibile d'essequir diverse statue senza aver precogitate diverse forme prima." This analogy is found in Plato's *Timaeus*, 29a; cf. *Phaedrus*, 249c, 249e, 250; cf. Plotinus, *Enneads*, Vol. I, Ennead i. 6. 2; 6. 8; 6. 9.

31. *Opere italiane*, I, 176: "Ill scopo e la causa finale, la qual si propone l'efficiente, è la perfezion dell'universo; la quale è che in diverse parti della materia tutte le forme abbiano attuale esistanza; nel qual fine tanto si deletta e si compiace l'intelletto, che mai si stanca suscitando tutte sorte di forme da la materia, come par che voglia ancora Empedocle."

32. Aristotle, *Physics*, ii. 7. 198a26.

33. *Ibid.*, 8. 199a30–33; cf. *ibid.*, 3. 194b33.

34. *Ibid.*, 3. 194b29.

35. *Ibid.*, 7. 198a19.

36. Plotinus, *Enneads*, II, 125 (Ennead ii. 8. 5).

37. During many discussions which took place while these passages were being analyzed, the writer experienced many difficulties with the attempt to show (*a*) how something can be logically distinct, without being really distinct, and (*b*) how neither of these concepts necessarily carries with it "separation." The best way to achieve understanding here is to give examples of each: thus a distinction can be made with regard to the arms, legs, and the head of a person, though this distinction does not imply separation; on the contrary, if you separate them, you are no longer dealing with a "person." Again, a real distinction can be made with regard to "principles of existence"; we can, for example, make a distinction between existence and essence. Lastly, we can deal with things which are actually separate in existence.

38. Cf. Tocco, *Le opere latine di Giordano Bruno*, p. 342: "I concetti fondamentali, onde muove, non sono più come nel De Umbris le idee, che ora non si dubita di chiamare fantastiche, ma invece le quattro cause aristoteliche. E se pur si fa menzione delle idee, non si considerano come transcendenti, ma invece quali cause formale, o per meglio dire quali norme direttive dell'anima del mondo nelle operazioni sue." This is born out by a comparison of passages in two separate works: cf. "Summa terminorum metaphysicorum," Intellectus seu idea, *Opera latine conscripta*, I, Part IV, 103: "Intelligentia ergo est divina quaedam vis, insita rebus omnibus cum actu cognitionis, qua omnia intelligunt, sentiunt, et quomo-

docunque cognoscunt"; and, "De umbris idearum," *ibid.*, pp. 43-44, Conceptio V, VI, VII. "Rerum formae sunt in ideis, sunt quodammodo in se ipsis, sunt in coelo; sunt in causis proximis seminalibus; sunt in proximis efficientibus, sunt individualiter in effectu, sunt in lumine, sunt in extrinsico sensu, sunt in intrinsico; modo suo."

39. *Opere italiane*, I, 175.

40. Plotinus, *Enneads*, IV, 4-5 (Ennead v. 1. 3); cf. *ibid.*, p. 16 (Ennead v. 2. 1); for the order of Hypostases see Ennead iii. 4. 1.

41. *Ibid.*, III, 12 (Ennead iv. 3. 4); cf. *ibid.*, p. 2 (Ennead iv. 2. 1); *ibid.*, pp. 156-157 (Ennead iv. 9. 4).

42. In point of fact, Bruno introduces a distinction between kinds of intellects at this stage which will prove to be the model for all such future resolutions. *Opere italiane*, I, 175: "There are three kinds of intellects: the divine, which is of all things; the mundane, which makes all things; and the other particular intellects, which become everything. Because it is necessary that between the extremes this intermediary [the world intellect] must be found, which is the true efficient cause, not only extrinsic, but also intrinsic of all natural things." In introducing this distinction of kinds, Bruno does not introduce a distinction of substance; as stated above, all these distinctions are distinctions of aspect only, whether that distinction be a formal one or one that has a "de facto" relationship.

43. Cf. chaps. vi, vii, and viii of "De immenso et innumberabilibus," *Opera latine conscripta*, Vol. I. "The soul of the world is the intrinsic principle and energy which animates it. It is the Spirit, and Life that penetrates all, is in all, and works throughout all."

44. *Opere italiane*, I, 177: "Ma, come uno efficient, separato dalla materia secondo l'essere dice che quello è cosa che viene di fuora, secondo la sua subsistenza, divisa dal composto." Cf. Aristotle, *De gen. anim.*, ii. 3. 736b28. Cf. St. Thomas Aquinas, *De unitate intellectus*, chap. iv (*In opuscoli e testi filosifici*, edited by Nardi [1916], pp. 30-33): "Anima autem intellectiva, cum habeat operationem sine corpore non est suum solum in concretione ad materiam; unde non potest dici, quod educatur de materia, sed magis quod est a principio extrinceco."

45. *Opere italiane*, I, 177: "Approvo quel che dite, perchè, se l'essere separata dal corpo all potenza intellettiva de l'anima nostra conviene, e lo aver raggione di causa efficiente, molto più si deve affirmare dell'anima del mondo." Cf. Gilson, *The Spirit of Medieval Philosophy*, pp. 177 ff.; in particular, the sections on Averroes and Avicenna. Cf. St. Thomas Aquinas, *Summa theologica*, I, 75 and 76.

46. *Opere italiane*, I, 177-181. In this section Bruno passes from the assertion that the world is "animated" to the consequent animation of all its parts. This is done by showing that there is nothing that has not the potentiality to become everything else, given the opportunity to realize that which it really is, as Substance, since the one substance, seen here as form or soul or intellect is the selfsame substance wherever it is. Hence: "Sia pur cosa quanto piccola e minima si voglia,

ha in se parte di sustanza spirituale . . . perchè spirto si trova in tutte le cose, e non e minimo corpusculo che non contegna cotal porzione in se, che non inanimi." Cf. Plotinus, *Enneads*, II, 31 (Ennead iii. 2. 16); *ibid.*, III, 92 (Ennead iv. 4, 36). "For this account allows grades of living within the whole, grades to some of which we deny life, only because they are not perceptibly self-moved, and in truth, all of these have a hidden life." Cf. Plato, *Timaeus*, 29–30.

47. *Opere italiane*, I, 194.

48. This throws some light on the statement made above—that is, "that everything is in everything." The statement itself is repeated by Bruno in "Sigillus sigillorum" (*Opera latine conscripta*, II, 3; and II, 196). "Unde cum anima ubique praesens existat illaque tota est in toto at in quacumque parte tota, ideo pro conditione materiae in quacumque re etiam exiqua et absicisa mundum, nedum mundi simulacrum valeas intueri, ut non temere omnia in omnibus dicere cum Anaxagora possimus." Cf. Cusanus, *De docta ignorantia*, edited by E. Hoffman and R. Klibansky (1932), I, Part II, 69–70, 73–74; Part I, 13, 25–27. Cf. Vansteenberghe, *Nicolaus of Cusa* (1920), pp. 291–292. For Cusanus, the universe is so constituted that each singular thing is the concrete expression of the totality; since the universe is in God as an effect is in its cause, and God is in the universe as a cause is in its effect, the universe is in every one of its parts, for every one of its parts is a restricted universe; hence, everything is in everything. For this, see especially Cusanus, *De docta ignorantia*, I, Part II, 88–89; *ibid.*, pp. 76–78. The comparison between Bruno and Cusanus is taken up in detail in Chapters II, III, and IV below.

49. *Opere italiane*, I, 183: "L'anima, dunque, del mondo è il principio formale constitutivo del'universo e di ciò che in quello si contiene."

50. *Ibid.*, p. 183: "La qual però, secondo la diversità delle disposizioni della materia, e secondo la facultà de principii materiali attivi e passivi, viene a produr diverse figurazioni, ed effettuar diverse facultadi."

51. *Ibid.*, p. 184: "Cossi, mutando questa forma sedie e vicissitudine, è impossibile che se annulle; perchè non e meno subsistente la sustanza spirituale che la materiale."

52. *Ibid.*, p. 184: "Dunque le formi esteriori sole si cangiano, e si annullano ancora, perchè non sono cose, ma de le cose; non sono sustanze, ma de le sustanze sono accidenti e circonstanze."

53. Cf. Mercati, "Il sommario del processo di Giordano Bruno," pp. 113 ff.

54. Plotinus, *Enneads*, III, 156–157 (Ennead iv. 9. 4); *ibid.*, pp. 12–13 (Ennead iv. 3. 4. 5).

55. Cf. McIntyre, *Giordano Bruno*, pp. 160–161: "In the De La Causa, Bruno maintains quite clearly the substantiality of the universal soul alone, the finite individual being merely one of the modes of its determination in matter."

56. Plotinus, *Enneads*, III, 92 (Ennead iv. 4. 36): "Similarly, the all could not have its huge life unless its every member had a life of its own." Cf. *ibid.*, III, 14 (Ennead iv. 3. 5): "Each soul is permanently a unity (a self) and yet all are, in their total, one being."

57. *Ibid.*, p. 153 (Ennead iv. 9. 1).

58. *Ibid.*, p. 150 (Ennead iv. 8. 6); *ibid.*, p. 11 (Ennead iv. 3. 3); *ibid.*, p. 14 (Ennead iv. 3. 5).

59. *Ibid.*, pp. 156–157 (Ennead iv. 9. 4).

60. *Opere italiane*, I, 186.

61. *Ibid.*, p. 187: "Oltre, in se invariabile, variabile poi per li soggetti e diversità di materie."

62. *Ibid.*, pp. 187–188. Cf. Mercati, "Il sommario del processo di Giordano Bruno," p. 113, No. 252; cf. *Opere italiane*, I, 400; cf. "De vinculis in genere," art. xv (*Opera latine conscripta*, p. 695).

63. In the closing passages of the dialogue, Bruno reviews his own stand with a brief summary of the synthesis that he has achieved so far. He calls attention to the fact that he agrees, in some measure, with Anaxagoras, Empedocles, Plato, and even Aristotle. He shows, however, that this is not to be taken as a mere addition of the views of others, but rather as a definite attempt to advance.

II THE MATERIAL PRINCIPLE AS POTENCY

1. *Opere italiane*, I, 197: "Ma, dopo aver più maturamente considerato, avendo risguardo a più cose, troviamo che e necessario conoscere nella natura doi geni di sustanza, l'ono che è forma, e l'altro che è materia." Cf. *ibid.*, p. 400. "E veramente è cosa necessaria, che, come possiamo ponere un principio materiale constante ed eterno, poniamo un similmente principio formale."

2. *Ibid.*, pp. 197–198.

3. *Ibid.*, p. 198. Cf. "De monade, numero et figura," *Opera latine conscripta*, I, Part II, 163 ff., 238 ff.

4. *Opere italiane*, I, 199: "Cossi la natura, a cui è simile l'arte, bisogna che de le sue operazioni abbia una materia; perchè non e possibile che sia agente alcuno, che se vuol far qualche cosa, non abbia di che farla; o, se vuol oprare, non abbia che oprare."

5. *Ibid.*: "E dunque una specie di soggetto, del qual, col quale, e nel quale effettua la sua operazione, ilsuo lavoro."

6. *Ibid.*, p. 200.

7. *Ibid.*, pp. 200 ff.

8. Plato, *Timaeus*, 52d; cf. 49a; 51a-b; 52b. Jowett translation.

9. Goheen, *Matter and Form in the De Ente et Essentia* (1940), pp. 48–49. The author draws an interesting comparison between Plato and St. Augustine which is especially relevant to our discussion. Speaking of St. Augustine, he says, "In the first place, unformed matter is so unformed as to be nearly nothing (informe prope nihil). This distinguishes it from the absolute nothingness from which in turn the unformed matter is made. The Scriptures, anticipating the fact that differences and difficulties would arise in the comprehension of an unformed matter which was yet something though derived from nothing, give the unformed nature of matter the name 'informitas' to indicate its state of existence between nothingness and formed matter. The unformed matter is unusual and so incon-

gruous as a concept that the senses are repulsed and the mind upset. It is a 'being which is not being' " (p. 49, n. 1). Augustine's language here suggests Plato's notion of the receptacle. As the Timaeus says, the nature of the receptacle can hardly be expressed and is understood only by a bastard sort of reason. Cf. *Timaeus*, 52b.

10. Aristotle, *Physics*, i. 7. 191a7–12; *ibid*., ii. 2. 194b16–26.

11. Aristotle, *De gen*., i. 4. 320a2–5.

12. Aristotle, *Physics*, i. 7. 191a7–12; *ibid*., 9. 192a25–33.

13. Plotinus, *Enneads*, II, 190 (Ennead ii. 4. 12): "It is grasped only by a mental process, though that not an act of the intellective mind, but a reasoning that finds no subject." Cf. *ibid*., p. 195 (Ennead ii. 4. 16): "Then matter is simply Alienism? No; it is merely that part of Alienism which stands in contradiction with the Authentic Existents which are Reason-Principles. So understand this non-existent has a certain measure of existence; for it is identical with Privation, which also is a thing standing in opposition to the things that exist in Reason. But must not Privation cease to have existence, when what has been lacking is present at last? By no means: the recipient of a state or character is not a state but the (Negation) or Privation of the state; and that into which determination enters is neither a determined object nor determination itself, but simply the wholly or partly *undetermined*." Cf. *ibid*., p. 194 (Ennead ii. 4. 15). "Matter is indetermination, and nothing else."

14. *Ibid*., p. 194. (Ennead ii. 4. 10). "This is Plato's reasoning where he says that matter is apprehended by a sort of spurious reasoning."

15. *Ibid*., p. 103 (Ennead iii. 7. 6).

16. Cf. *Opere italiane*, I, 400: "Il summo agente potente fare il tutto con il possible esser fatto il tutto coincideno in uno."

17. *Ibid*., pp. 205–206: "E veramente è cosa necessaria, che, come possiamo ponere un principio materiale constante ed eterno, poniamo un similmente principio formale."

18. *Ibid*., p. 206.

19. *Ibid*., pp. 206 ff.

20. *Ibid*., p. 207: "Onde diciamo in questo corpo tre cose: prima l'intelletto universale, indito nelle cose; secondo, l'anima vivaficatrice del tutto; terzo, il soggetto."

21. *Ibid*., p. 211: "Certo, questo principio, che è detto materia, può essere considerato in doi modi: prima, come una potenza, secondo, come un soggetto."

22. *Ibid*., p. 211: "E. cossi non è cosa di cui si può dir l'essere, della quale non si dica il posser essere . . . onde, se sempre, e stata la potenza di fare, di produre, di creare, sempre e stata la potenza di esserfatto, produtto, e creato; perchè la una potenza implica l'altra."

23. Cusanus, *De docta ignorantia*, I, Part II, 8, 84–89—esp. p. 88. *Ibid*., pp. 64–65. Cf. Vansteenberghe, *Nicolaus of Cusa*, p. 307.

24. *Opere italiane*, I, 212.

25. *Ibid.* Cf. Cusanus, "De possest," *Opera* (1565), p. 251: "Absolute potency and absolute act are coeternal."

26. *Opere italiane*, I, 212. Cf. Cusanus, "De possest," *Opera*, p. 251: "Ubique est magnus, sed sic magnus, quod magnitudo, quae est omne id, quod esse potest."

27. *Opere italiane*, I, 212.

28. *Ibid.*, p. 213: "L'universo è tutto quel che può essere, secondo un modo esplicato, disperso, distinto. Il principio suo è unitamente e indifferentemente; perchè tutto è tutto e il medesmo simplicissimamente, senza differenza e distinzione." Cf. Cusanus, *De docta ignorantia*, I, Part II, 69–72; *ibid.*, pp. 72–75; Cusanus employed two similar terms, "maximum absolutum" and "maximum contractum," for Bruno's "complicatio" and "explicatio." For Cusanus, God's relation to the world is that of unity to multiplicity, of simplicity to extension, of eternity to time, of rest to movement; God is the unity of simplicity, identity, eternity, and rest.

29. Cf. "De l'infinito," *Opere italiane*, I, 291. Here Bruno expressly distinguishes the infinity of God from the infinity of the universe: the first, he calls "intensive," the second he calls "extensive." "Io dico l'universo tutto infinito, perchè non ha margine, termine, ne superficie; dico L'universo non essere totalmente infinito, perchè ciascuna parte, che di quello possiamo prendere, e finita, e de'mondi innumerabili, che contiene, ciascuno è finito. Io dico Dio tutto infinito, perchè da se esclude ogni suo attributo e uno e infinito; e io dico Dio totalmente infinito, perchè tutto lui è in tutto il mondo e in ciascuna sua parte infinitamente e totalmente." For Cusanus, the infinite can be applied to the universe only in the sense that the universe is the greatest of created things; the universe is not limited by anything, but itself limits all things. It is thus a "maximum contractum," while God is the "maximum absolutum." The universe is, therefore, for Cusanus, a relative or a privative infinite. Cf. *De docta ignorantia*, I, Part II, 75. Cf. also Cusanus, "De possest," *Opera*, p. 251: "Volo dicere quod omnia *il*la complicite in Deo sint Deus, sicut explicit in creatura mundi sunt mundus."

30. *Opere italiane*, I, 216: "Onde cossi de l'universo sia un principio che medesmo se intenda, non più distintamente materiale e formale, che possa inferirse dalla similitudine del predetto, potenza absoluta e atto." Cf. "De vinculis in genere," Art. xv (*Opera latine conscripta*, III, 695). "Et divinum ergo quoddam est materia, sicut et divinum quoddam existimatur esse forma, quae aut nihil est aut materiae quiddam est."

31. Mercati, *Il sommario del processo di Giordano Bruno*, p. 80; *ibid.*, pp. 113–114.

32. *Opere italiane*, I, 400. Cf. Namer, *Les aspects de Dieu dans la philosophie de Giordano Bruno*, p. 147: "There are two infinites. The first is 'intensive,' the complicative infinity of God; the second is 'extensive,' the explicative infinity of the universe; but substance is one and identical. God and the universe are two points of view of the same reality."

Bruno closes the dialogue with an example of the simultaneous presence of God, which is reminiscent of Cusanus. It is based upon the fact that God, since he has the capacity to be everywhere simultaneously, realizes that capacity because it is one with his actuality. Cf. Cusanus "De possest," *Opera*, p. 252.

III THE MATERIAL PRINCIPLE AS SUBJECT

1. *Opere italiane*, I, 224. "E certamente non si può negare, che, si come ogni sensibile presuppone il soggetto della sensibilità, cossi ogni intelligibile il soggetto della intelligibilità."

2. *Ibid.*

3. Cf. St. Thomas Aquinas, *De ente et essentia*, translated by G. Leckie (1937), chap. iv, pp. 21ff. "Now it remains to see through what mode essence exists in separate substances, namely, in the soul, in intelligences, and in the first cause. But although all grant the simplicity of the first cause, yet certain ones strive to introduce a composition of form and matter in intelligences and in the soul. The author of this position appears to have been Avicebron, the writer of the book, 'Fons Vitae.' "

In St. Thomas, God's essence is identical with His Existence; in created intellectual substances, essence is other than existence, though their essences are without matter; in substances composed of matter and form, the essence is both. The hierarchy of being, for St. Thomas, proceeds upward from the pure potentiality of matter, which is actualized by form, to the level of form (this time itself potential to existence) plus existence, without matter, to the level of pure act.

4. *Opere italiane*, I, 226 ff.

5. Plotinus, *Enneads*, II, 178–181 (Ennead ii. 4. 1–5).

6. *Ibid.*, p. 178 (Ennead ii. 4. 1).

7. *Ibid.* (Ennead ii. 4. 2.).

8. *Ibid.*, p. 179 (Ennead ii. 4. 3).

9. *Ibid.*

10. *Ibid.*

11. *Ibid.*, p. 181 (Ennead ii. 4. 5).

12. *Opere italiane*, I, 226. Cf. Tocco, *Le opere latine di Giordano Bruno*, p. 342: "Queste differenze (vole a dire corporale e incorporale sustanza) si riducone alla potenza di medesimo geno."

13. *Opere italiane*, I, 227.

14. *Ibid.*: "Cossi ad una potenza attiva, tanto di cose corporali, quanto di cose incorporee, over ad un essere, tanto corporeo, quanto incorporeo, corrisponde una potenza passiva, tanto corporea, quanto incorporea, e un posser esser, tanto corporeo, quanto incorporeo."

15. *Ibid.*, p. 228: "Quella materia, per esser attualmente tutto quel che può essere, ha tutte le misure, ha tutte le specie di figure e di dimensioni; e perchè le ave tutte, non ne ha nessuna; perchè quello ch'e tante cose diverse, bisogna che non sia alcuna di quelle particolari."

16. *Ibid.*, p. 229.

17. *Ibid.*, p. 230.

18. *Ibid.*, p. 234.

19. *Ibid.*, p. 237: "What can a corruptible thing give to an eternal thing? An imperfect thing, like the form of sensible things which is always in movement—what can this give to the perfect?"

20. Cf. "De immenso et innumerabilibus," chap. viii (*Opere latine conscripta*, I, Part II, 33 ff.).

21. *Opere italiane*, I, 245–247. Cf. *ibid.*, p. 400.

22. *Ibid.*, p. 400: "E il summo agente e potente fare il tutto con il possible esser fatto il tutto coincideno in uno." Cf. Mercati, "Il sommario del processo di Giordano Bruno," p. 113.

23. Cf. Tocco, *Le opere latine di Giordano Bruno*, p. 342. Cf. McIntyre, *Giordano Bruno*, p. 339: "In Bruno, matter and form are not two principles; they are one and the same thing." Cf. Namer, *Les aspects de Dieu dans la philosophie de Giordano Bruno*, pp. 145–147. In the previous chapter, the difference between Bruno and Plato, Aristotle, and Plotinus served to clarify the text. Here, the same subject matter calls forth the same comparisons; however, to these can now be added the following: Against Plato, Bruno argues that it is not necessary to return to the fantasy of the theory of Ideas; Ideas are within matter, and not separate from them; cf. "De immenso et innumerabilibus," chap. viii. (*Opera latine conscripta*, Vol. I). For Aristotle, matter "desires" its perfection in act; for Bruno, matter is perfect in itself; it cannot desire that which it already possesses; cf. Aristotle, *Physics*, i. 9. 192a25ff.; *ibid.*, 7. 191a7ff. Finally, where Plotinus would declare the One to be devoid of any form of determination, and consequently declare it to be transcendent, Bruno identifies "Unity," with "the all that can be"; and where the One in Plotinus engenders substances inferior to itself, the One of Bruno cannot engender other substances, precisely because it is the only Substance; cf. Plotinus, *Enneads*, Vol. V (Ennead vi. 7); *ibid.* (Ennead vi. 9); *ibid.*, Vol. III (Ennead iv. 9. 4); *ibid.* (Ennead iv. 8. 6); *ibid.* (Ennead iv. 4. 36).

IV SUBSTANCE AND INFINITY

1. *Opere italiane*, I, 239: "E dunque l'universo uno, infinito, immobile. Una, dico, è la possibilità assoluta, un l'atto, una la forma, o la anima, una la materia, o corpo, una la cosa, uno lo ente, uno il massimo e ottimo; il quale non deve posser essere compreso; e però infinibile e interminabile, e per tanto infinito e interminato, e per conseguenza immobile."

2. *Ibid.*, p. 239. Cf. Mercati, "Il sommario de processo di Giordano Bruno," p. 114, No. 255.

3. *Opere italiane*, I, 240: "Perchè è il tutto indifferente, e però è uno, l'universo è uno."

4. *Ibid.*, p. 240.

5. *Ibid.*, pp. 240–241: "Alla proporzione, similitudine, unione e identità del'infinito non più ti accosti con essere uomo che formica, una stella che un uomo; e però nell'infinito queste cose sono indifferenti."

6. *Ibid.*, p. 241: "Dunque, l'individuo non e differente dal dividuo, il simplicissimo da l'infinito, il centro da la circonferenza."

7. *Ibid.*, p. 242.

8. *Ibid.* Cf. *ibid.*, p. 291; cf. Cusanus, "*De docta ignorantia*," I, Part II, 68–72, 72-75; cf. Cusanus, "De possest," *Opera*, p. 252.

9 *Opere italiane*, I, 244.

10. *Ibid.*, p. 246.

11. Bruno declares, at this point, "that all contradictory propositions are true"; this is not to be construed as meaning the "principle of the coincidence of opposites and contraries"; this statement refers to the fact that since substance is one wherever it is, every proposition enunciated concerning it is true. Cf. "If we start with the proposition that all contradictions are true we arrive at the identity of being"; Aristotle, *Metaphysics*, iii. 4. 1007a17: "If all contradictory affirmations are true, relative to the same thing, all things would be a single thing, a ship, a wall, a man." *Ibid.*, 1007b20–25: "The opinion of such men does not merit serious attention. For, in fact, they do not say anything." *Ibid.*, 1007b25: "The object of their discussion is not being, but the indeterminate. They are indeed as vegetables." Bruno uses the word "true" in a different sense from Aristotle. For Bruno, a statement can be called true as long as it refers to that which is. And in this sense, contradictory statements must be true, since if they did not both refer to that which is, and hence possess truth in Bruno's sense, they would not be contradictory, since contradiction must coincide in a common subject.

12. Plotinus, *Enneads*, V, 114 (Ennead vi. 4. 7).

13. *Opere italiane*, I, 250.

14. Plotinus, *Enneads*, V, 238 (Ennead vi. 9. 1).

15. *Ibid.*, II, 132 (Ennead iii. 8. 9); cf. *ibid.*, p. 130 (Ennead iii. 8. 8).

16. *Ibid.*, V, 241 (Ennead vi. 9. 3).

17. De Deo seu mente, "Summa terminorum metaphysicorum," *Opera latine conscripta*, I, Part IV, 73–78: "God is absolutely omnipresent."

18. Plotinus, *Enneads*, V, 111 (Ennead vi. 4. 4); cf. *ibid.*, III, 150 (Ennead iv. 8. 6); *ibid.*, p. 157 (Ennead iv. 9. 4.).

19. *Opere italiane*, I, 242–243.

20. *Ibid.*, p. 247.

21. Cusanus, *De docta ignorantia*, I, Part II, 69–72 (especially p. 71); *ibid.*, pp. 73–74; cf. *ibid.*, Part I, p. 27. For Cusanus, God is the "maximum absolutum," and at the same time the "minimum absolutum"; what is opposed and contrary in particular things, is in God identical. To arrive at the conception of the principle of opposites and contraries, Cusanus employs the geometrical examples which Bruno has taken as his own—*ibid.*, Part I, p. 25: "Si esset linea infinita, illa esset recta, illa esset triangulus, illa esset circulus, et esset sphaera." Cf. Summa terminorum metaphysicorum," *Opera latine conscripta*, I, Part IV, 83–84.

22. Cf. "De immenso et innumerabilibus," *Opera latine conscripta*, I, Part II, 12. cf. Mercati, "Il sommario del processo di Giordano Bruno," p. 80; cf. "La cena de le ceneri," *Opere italiane*, I, 96; *ibid.*, pp. 297 ff.

23. *Opere italiane*, I, 246: "La qual distinzione e sglomeramento non viene a produre altra e nuova sustanza, ma viene a ponere in atto e compimento certe qualitadi, differenze, accidenti, e ordini, circa quella sustanza." Cf. "De monade, numero et figura," *Opera latine conscripta*, I, Part II, 216 ff.

24. *Opere italiane*, I, 245.

25. Cf. "Summa terminorum metaphysicorum," *Opera latine conscripta*, I, Part

IV, 73: "Deus ergo est substantia universalis in essendo, qua omnia sunt, essentia omnis essentiae fons, qua quidquid est, intima omni enti magis quam sua forma et sua natura unicuique esse possit. Sicut enim natura est unicuique fundamentum entitatis, ita profundius naturae uniuscuiusque fundamentum est Deus. Propterea bene dicitur in quo vivimus, vegetamur, et sumus; quia est vitae vita, vegetationis vegetatio, entitatis essentia." Since the universe is the infinite image of the infinite substance, and the "effect" of that infinite cause, the things of the universe must bear a relation to one another, which resembles the relation of the "parts" of substance to one another. Since substance is one wherever it is, and we can say that substance is in the parts, we can also say that the parts are universally related; and it is likewise safe to say that every thing is in everything, since everything is a contracted expression of substance.

V CONCERNING THE INFINITE UNIVERSE AND WORLDS

1. *Opere italiane*, I, 290.
2. Cf. Tocco, *Le opere latine di Giordano Bruno*, pp. 211–236. In these pages, Tocco compares "De l'infinito" with "De immenso," by setting side by side parallel passages from both; see especially pp. 317 ff.
3. *Opere italiane*, I, 291.
4. *Ibid.*, pp. 291 ff. The terms "complicatio" and "explicatio" are especially helpful here; the intensive infinite is all things as enfolded, while the extensive infinite is all things as unfolded; the universe is the unfolded image and aspect of the intensive infinite which is being reflected in the universe.
5. *Ibid.*, p. 281.
6. Aristotle, *Physics*, iv. 4. 212a20; cf. *Opere italiane*, I, 281.
7. *Opere italiane*, I, 281: "Il convesso del primo cielo e loco universale; e quello, come primo continente, non è in altro continente; perchè, il loco non è altro che superficie ed estremità di corpo continente; onde chi non ha corpo continente, non ha loco."
8. *Ibid.*, p. 282.
9. Aristotle, *Physics*, iv. 5. 212b11–12.
10. *Ibid.*, 212b8–22.
11. *Ibid.*, 212b14–18.
12. *Ibid.*, 1. 209a23–26.
13. *Ibid.*, 209a26.
14. *Ibid.*, 3. 210b8–10.
15. *Ibid.*, 210b23.
16. *Opere italiane*, I, 282. The difference, of course, which Bruno does not seem to care to emphasize, is that he is using the term "outside" in a spatial sense, while Aristotle is not.
17. "De immenso et innumerabilibus," *Opera latine conscripta*, I, Part I, 222; cf. Cusanus, "De possest," *Opera*, p. 252.
18. "De immenso et innumerabilibus," *Opera latine conscripta*, I, Part I, p. 227.
19. *Opere italiane*, I, 264.

20. Cf. Lucretius, *De rerum natura*, Bk. I, ll. 998–1001, 1006–1007. Cf. ll. 973–989:

> . . . It matters nothing where thou post thyself
> In whatever regions of the same
> Even any place a man has set him down
> Still leaves about him the unbounded all
> Outward in all directions.

Cf. *Opere italiane*, I, 264.

21. *Opere italiane*, I, 284: "Perchè non possiamo fuggire il vacuo, se vogliamo ponere l'universo finito."

22. *Ibid.*, p. 264: "Citra il mondo, dunque, è indifferente questo spacio da quello; dunque l'attitudine ch'ha questo spacio, ha quello." Cf. "De immenso et innumerabilibus," *Opera latine conscripta*, I, Part II, 9.

23. Aristotle, *Physics*, iv. 8. 215a8–11: "The void seems to be a non existent and a privation of being."

24. "De immenso et innumerabilibus," *Opera latine conscripta*, I, Part I, 232.

25. *Opere italiane*, I, 290 ff. Cf. "De immenso et innumerabilibus," *Opera latine conscripta*, I, Part I, 235.

26. *Opere italiane*, I, 290 ff. Cf. *ibid.*, pp. 399 ff.: "In esso sono infiniti mondi simili a questo e non differenti in geno da questo."

27. Bruno is not attacking those arguments of the *Physics* wherein Aristotle opposes the existence of a vacuum within the world. These arguments are found in *Physics*, iv. 7–9—in particular, iv. 7. 214a16–29: "Since we have determined the nature of place, and void must, if it exists, be place deprived of body, and we have made it plain both in what sense it (place) exists and in what sense it does not, it is plain that on this showing, void does not exist either separated or unseparated; for the void is meant to be, not body, but an interval of body." Cf. *ibid.*, iv. 8. 214b–215a.

28. *Ibid.*, 5. 212a32; cf. *ibid.*, 6. 213b32–214a. Cf. *ibid.*, 8. 214b18–20: "Again if void is a sort of place deprived of body, where there is a void, where will a body placed in it move to?"

VI INTENSIVE AND EXTENSIVE INFINITY

1. These arguments comprise the whole of the first dialogue of "De l'infinito, universo e mondi" (*Opere italiane*, I, 259–414).

2. Aristotle, *De coelo*, i. 5. 271b17–277a8. Cf. *Opere italiane*, I, 305–306.

3. Aristotle, *De coelo*, i. 5. 271b17.

4. *Opere italiane*, I, 307. Cf. Aristotle, *De coelo*, i. 6. 273a1–22: "If the center is determinate, so is the circumference, and the intermediate portion as well, and thus bodies moving in it are necessarily determinate."

5. Aristotle, *De coelo*, i. 5. 272b25–273a5.

6. Bruno returns to the Aristotelian treatment of the other kinds of simple body in what follows.

7. *Opere italiane*, I, 310: "La terra, dunque, non è assolutamente in mezzo de l'universo ma al riguardo di questa nostra raggione."

8. Cf. Aristotle, *De coelo*, i. 1–3. Aristotle had maintained that there are types of moving bodies which are heavy, light, and neutral; earth and water are heavy; air and fire are light; and the quintessence is neutral. All three follow from the Aristotelian conception of the heavenly circumference with its fixed center; thus, the heavy belongs to the center, the light to the space between the center and the circumference, and the neutral to the circumference. Even if Bruno were to accept the fact that there might be many centers, if it is held that there are as many centers as there are worlds, this would still not entail that the parts of any one move to any centers than their own; cf. *ibid.*, and i. 8. For Aristotle, if it is assumed that the elements of our world are the same as those of the other worlds, then it must also be assumed that the elements, in their natural movements, would be moving to and from a center, and to and from a circumference. Cf. *Opere italiane*, I, 347.

9. *Opere italiane*, I, 311–312.

10. Cf. Aristotle, *De coelo*, i. 7. 274a30–b10; cf. "De immenso et innumerabilibus," *Opera latine conscripta*, I, Part I, 267.

11. *Opere italiane*, I, 314. Cf. Lucretius, *De rerum natura*, Bk. II, ll. 505–557.

12. See Chapter IV above.

13. The different "kinds" of atomism in Bruno are discussed in what immediately follows.

14. *Opere italiane*, I, 315.

15. *Ibid.*, p. 316.

16. *Ibid.*, p. 317.

17. *Ibid.*

18. *Ibid.*

19. *Ibid.*

20. *Ibid.*, p. 319: "Il corpo dunque infinito, secondo noi, non è mobile, ne in potenza, ne in atto; e non è grave ne lieve in potenza, ne in atto." Cf. Aristotle, *De coelo*, i. 7. 274b30–34: "Moreover, in general, it is impossible that the infinite should move at all. If it did, it would move either naturally or by constraint; and if by constraint, it possesses also a natural motion, that is to say, there is another place, infinite like itself, to which it will move. But that is impossible."

21. Aristotle, *De coelo*, i. 7. 275b5–7; *Opere italiane*, I, 319 ff. Cf. "De immenso et innumerabilibus," *Opera latine conscripta*, I, Part II, 6.

22. *Opere italiane*, I, 326. Bruno has reversed the Aristotelian position. Action and passion are impossible with regard to the infinite, not, as Aristotle says, because action and passion are confined to the finite, but because the action and passion involved in the finite cannot be attributed to the infinite which has no parts. The fact that action and passion are of the sensible does not argue against the actual existence of the infinite; the sensible action must be referred only to the discrete and the separate where the infinite is not exerting its total power; besides, the infinite is not agent or patient, but it is both agent and patient. Cf. Aristotle, *De coelo*, i. 7; and *Physics*, iii. 5. 2-4b22–35: "Nor can the infinite body be one and simple, whether it is as some hold, a thing over and above the elements (from which they generate the elements) or is not thus qualified. We

must consider the former alternative; for there are some people, who make this the infinite, and not air or water, in order that the other elements may not be annihilated by the element which is infinite." Cf. *Opere italiane*, I, 239 ff.

VII THE INFINITE UNIVERSE

1. *Opere italiane*, I, 364–365. Cf. Aristotle, *De coelo*, i. 8. 276a22-b21.
2. Cf. Aristotle, *Physics*, iii. 5. 205a30–36; *Opere italiane*, I, 314; cf. Aristotle, *De coelo*, i. 7.
3. *Opere italiane*, I, 336–367.
4. Aristotle, *De coelo*, i. 8. 276b21–25; *Opere italiane*, I, 370.
5. Aristotle, *De coelo*, i. 8. 276a18–19.
6. *Ibid.*, 276b28–30.
7. *Ibid.*, 276b30–31.
8. *Ibid.*, 276b32–277a2.
9. *Ibid.*, 277a5–10.
10. In the section comprised in pp. 373–378, inclusive, of *Opere italiane*, Vol. I, Bruno takes up points which are in essence a repetition of what has preceded; that is, he admits that motion is from opposite to opposite; he admits that motion is determinate; but he does not admit that from this it follows that the universe is finite nor that the world is one.
11. "De immenso et innumberabilibus," *Opera latine conscripta*, I, Part II, 274.

VIII SUBSTANCE, UNIVERSE, AND THE INFINITE

1. *Opere italiane*, I, 390. Cf. Aristotle, *De ceolo*, i. 9. 278a27–28.
2. *Opere italiane*, I, 390–391.
3. *Ibid.*, pp. 390 ff. Pages 387–389 in this fifth dialogue are taken up in the introduction of a new interlocutor who represents alien interests, and whom Bruno attempts to impress by a summary of some of his new revelations in the field of "the philosophy of nature." This impression, as we shall see, is a favorable one, for the reason that Bruno makes it so.
4. Aristotle, *De coelo*, i. 9. 278b29–279a8. *Opere italiane*, I, 390–391.
5. Aristotle, *De coelo*, i. 9. 279a12–19; cf. *Opere italiane*, I, 390 ff.
6. *Opere italiane*, I, 399: "Cotal spacio, lo diciamo infinito, perchè non è raggione, convenienza, possibilità, senso, o natura, che debba finirlo; in esso sono infiniti mondi simili a questo, e non differenti in geno da questo; perchè non e raggione, nè difetto di facultà naturale, dico tanto potenza passiva, quanto attiva, per la quale, come in questo spacio circa noi ne sono, medesimamente non ne sieno in tutto l'altro spacio, che di natura non è differente e altro da questo. Cf. Mercati, "Il sommario del processo di Giordano Bruno," pp. 113–114.
7. *Opere italiane*, I, 396 ff.
8. Ibid., p. 399. See note 1 above.
9. Aristotle, *De coelo*, iii. 2. 300b31–301a24; *Opere italiane*, I, 391 ff.
10. Aristotle, *Physics*, viii. 5. 256a4–257b26. Cf. *ibid.*, 258a1–2: "Therefore in

the whole of the thing we may distinguish that which imparts motion without itself being moved and that which is moved; for only in this way is it possible for a thing to be self-moved."

11. Aristotle, *Metaphysics*, xii. 1072a–1074a31; cf. Aristotle, *Physics*, viii. 5. 256a4 ff.

12. Cf. *Opere italiane*, I, 400; cf. "De immenso et innumerabilibus," *Opera latine conscripta*, I, Part II, 13.

IX CONCLUSION—THE CONCEPT OF INFINITY

1. *Opere italiane*, I, 239.
2. *Ibid.*
3. *Ibid.*
4. *Ibid.*
5. *Ibid.*
6. *Ibid.*
7. *Ibid.*
8. *Ibid.*, p. 240.
9. *Ibid.*
10. *Ibid.*
11. *Ibid.*, p. 241.
12. *Ibid.*, p. 242.
13. *Ibid.*, p. 251; cf. *ibid.*, pp. 240 ff.
14. *Ibid.*, p. 252. Cf. Cusanus, *De docta ignorantia*, I, Part I, 26.
15. *Opere italiane*, I, 254. Cf. *ibid.*, p. 253. It must be kept in mind that Bruno does not believe that an infinite line, or an infinite triangle, or an infinite angle, actually exists: "Therefore, if you take an infinite triangle—I do not mean really or absolutely, because the infinite has no figure, but an infinite triangle by supposition." Cf. Cusanus, *De docta ignorantia*, I, Part I 25–30.
16. *Opere italiane*, I, 291. Cf. *ibid.*, pp. 256–257: "Quella unità è tutto, la quale non è esplicata, non è sotto distrubuzione di numero, e tal singularità che tu intendereste forse, ma che è complicante e comprendente."
17. The reason that this is given such importance is that there remains always the tendency to revert to one's own terms, with the result that the whole edifice crumbles with a crumbling of the "distinctive" bricks. Bruno must be understood in line with *his* definitions and *his* terms; we cannot superimpose our own and proceed to call for an explanation that is not relevant.
18. *Opere italiane*, I, 291. The distinction between the two kinds of infinity is here applied to the concept of Substance; in point of fact, we can take as an example of the extensive infinite the sum of the particulars, because in the definitions of the "infinites," the major difference is based on the fact that in the universe each particular is encountered as a finite.
19. *Ibid.*, p. 243: "Ogni produzione, di qualsivoglia sorte che la sia, è una alterazione, rimanendo la sustanza medesima."
20. *Ibid.*, p. 244.

21. *Ibid.*, p. 245.

22. *Ibid.*, p. 246.

23. *Ibid.*, pp. 298–299.

24. *Ibid.*, p. 214; cf. "Summa terminorum metaphysicorum," *Opera latine conscripta*, I, Part IV, 94.

25. *Opere italiane*, I, 214; cf. *ibid.*, p. 400.

26. "Summa terminorum metaphysicorum," *Opera latine conscripta*, I, Part IV, 83: "Ut infra utrumque terminum vel utriusque terminus sit unus, principium huius et finis illius. Ubi enim est motus et alteratio, perpetuo extremum unius est initio alterius contrari." Cf. "Eroici furori," *Opere italiane*, II, 324: "Il fine d'un contrario e principio de l'altro, e l'estremo de l'uno è cominciamento de l'altro."

27. "Eroici furori," *Opere italiane*, II, 324-325; cf. *Opere italiane*, I, 255. Cf. "Summa terminorum metaphysicorum," *Opera latine conscripta*, I, Part IV, 83.

28. *Opere italiane*, I, 255–256: "Profundo magia e saper trar il contrario, dopo aver trovato il punto de l'unione." Cf. *ibid.*, II, 325: "La contrarietade è massime là dove è l'estremo; la contrarietà maggiore è la più vicina all'estremo; la minima o nulla è nel mezzo, dove gli contrarli convegnono e son uno e indifferente." Cf. *ibid.*, I, 256.

29. Cf. Nicholas of Cusa, *The Vision of God*, Salter translation (1928), p. 60: "Since Thine end is Thine essence, the essence of the end is not determined or ended in any place other than the end, but in itself. The end then which is its own end is infinite, and every end which is not its own end is a finite end." Cf. *ibid.*, p. 62: "Absolute infinity includeth and containeth all things. Did not infinity include in itself all being, it were not infinity. If it were not infinity, then neither would the finite exist, nor aught alien or different, since these cannot exist without otherness of ends or limits."

30. "De umbris idearum," Conceptio x, *Opera latine conscripta*, II, Part I, 45: "Hae res cum profluunt aliae ab aliis, diversae a diversis; in innumerum multiplicantur ut eas nisi qui numerat multidinem stellarum, non derminet: Cum vero refluunt uniuntur usque ad ipsam unitatem quae unitatem omnium fons est." Cf. Nicholas of Cusa, *The Vision of God*, p. 60; cf. "Intellectus seu idea," *Opera latine conscripta*, I, Part IV, 112.

31. "De umbris idearum," *Opera latine conscripta*, II, Part I, Conceptio xii: "Virtutes enim quae versus materiam explicantur: versus actum primum uniuntur et complicantur . . . . Item et in prima mente unam esse rerum omnium ideam, illuminando igitur, vivificando et uniendo, est quod te superioribus agentibus conformans, in conceptionem et retentionem specierum efferaris." Cf. *Opere italiane*, I, 247. Cf. also Nicholas of Cusa, *The Vision of God*, pp. 60–62.

32. *Opere italiane*, I, 247: "Talmente l'uno ente summo, nel quale è indifferente l'atto dalla potenza, il quale può essere tutto assolutamente, ed e tutto quello che può essere, è complicatamente uno, immenso, infinito, che comprende tutto lo essere, ed è esplicatamente in questi corpi sensibili e in la distinta potenza e atto che veggiamo in essi." Cf. *ibid.*, pp. 251–256.

33. *Ibid.*, p. 253.

34. *Ibid.*, p. 241: "Or, se tutte queste cose particulari ne l'infinito non sono

altro e altro, non sono differenti, non sono specie, per necessaria consequenza non sono numero; dunque, l'universo è ancor uno immobile." Cf. Nicholas of Cusa, *The Vision of God*, p. 62: "Absolute infinity includeth and containeth all things. . . . If it were not infinity, then neither would the finite exist, nor aught alien or different, since these cannot exist without otherness of ends and limits. . . . Infinity, accordingly, existeth, and enfoldeth all things, and naught can exist outside it, hence naught is alien to it or differing from it. Thus infinity is alike all things and no one of them all." Cf. also *ibid.*, pp. 61–62. Cf. also Cusanus, *De docta ignorantia*, I, Part II, 3: "Unitas infinita est omnium complicatio." *Ibid.*: "Excedit mentem nostram modus complicationis et explicationis."

35. *Opere italiane*, I, 291: "Io dico l'universo tutto infinito, perchè non ha margine, termine, ne superficie; io dico l'universo non essere totalmente infinito, perchè ciascuna parte, che di quello possiamo prendere, è finita, e de'mondi innumerabili, che contiene, ciascuno è finito. Io dico Dio tutto infinito, perchè da sè esclude ogni termine, e ogni suo attributo è uno e infinito; e dico Dio totalmente infinito, perchè tutto lui è in tutto il mondo, e ciascuna sua parte infinitamente e totalmente." Cf. Cusanus, *De docta ignorantia*, I, Part II, 3, 4. See section above, "The Conception of Infinity."

36. *Opere italiane*, I, 173.

37. *Ibid.*, p. 175.

38. See Chapter I above.

39. See Chapter I for a comparison with Aristotle and Plotinus.

40. *Opere italiane*, I, 251.

41. *Ibid.*, p. 250: "La prima intelligenza in una idea perfettissimamente comprende it tutto; la divina mente e la unità assoluta, senza alcuna specie, è ella medesima lo che intende e lo che è inteso."

42. *Ibid.*, II, 13: "Dove dunque era l'orsa, per raggion del luogo, per essere parte più eminente del cielo, si prepone la Verità; la quale è più alta e degna de tutte cose, anzi la prima, ultima e mezza; perchè ella empie il campo de l'Entità, Necessità, Bontà, Principio, Mezzo, Fine, Perfezione; si concepe negli campi contemplativi metaphysico, fisico, morale, logicale." Cf. *ibid.*, I, 76-77.

43. *Ibid.*, I, 247: "Prima dunque, voglio che notiate essere una e medesima scala, per la quale la natura descende alla produzion de le cose, e l'intelletto ascende alla cognizion di quelle."

44. For a detailed treatment of the relation between particular intellects and the universal intellect, see Chapter I. While Bruno introduces a distinction of kinds, he does not introduce a distinction of substance.

45. *Opere italiane*, I, 248.

46. *Ibid.*, pp. 249–250.

47. *Ibid.*, p. 250: "Cossi dunque, montando noi alla perfetta cognizione, andiamo complicando la moltitudine; come, descendendosi alla produzione delle cose, si va esplicando la unità. Il descenso lo ascenso è da questi a quello." See Chapter IV above.

48. "De immenso et innumerabilibus," *Opera latine conscripta*, I, Part II, 1–32.

49. See Chapter I, summary.

BIBLIOGRAPHY

EDITIONS OF BRUNO'S WORKS

For a complete bibliography of the works of Bruno, see V. Salvestrini, Bibliografia delle opere di Giordano Bruno, e degli scritta ad esso attinenti, Pisa, 1926.

1830, Leipzig. Opere di Giordano Bruno, ora per la prima volta raccolte e publicate da Adolfo Wagner. 2 vols. in 1.

1834, Stuttgart, Paris, and London. Jordani Bruni scripta, quae latine confecit, omnia, collegit, praefatione instruxit, mendisque expurgavit innumeris A. Fr. Gfrorer.

1868, Berlin. Jordanus Brunus Nolanus, De umbris idearum, edita nova, curavit Salvator Tugini.

1879–1891, Naples and Florence. Jordani Bruni Nolani Opera latine conscripta, publicis sumptibus edita (recensebant F. Fiorentino, F. Tocco, H. Vitelli, V. Imbriani, et C. M. Tallarigo). 3 vols. in 8 parts.

Vol. I is in four parts. Part I contains (1) Oratio valedictoria (Wittenberg, 1588), (2) Oratio consolatoria (Wittenberg, 1588), (3) Acrotismus camoeracensis (Wittenberg, 1588), (4) Books I–III of De immenso et innumerabilibus (Frankfort, 1591). Part II contains (1) Books IV–VIII of De immenso et innumerabilibus (Frankfort, 1591) and (2) De monade, numero et figura (Frankfort, 1591). Part III contains (1) Articuli adversus mathematicos (Prague, 1588) and (2) De triplici minimo et mensura (Frankfort, 1591). Part IV contains (1) Summa terminorum metaphysicorum (Turin, 1595), (2) Figuratio physici auditus Aristotelis (Paris, 1586), (3) Mordentius et de mordentii circino (Paris, 1586).

Vol. II is in three parts. Part I Contains (1) De umbris idearum (Paris, 1582), (2) Ars memoriae (Paris, 1582), (3) Cantus circaeus (Paris, 1582). Part II contains (1) De architectura lulliana, (2) Ars reminiscendi, trigiunta sigilli . . . . sigillus sigillorum (London, 1583), (3) Centum et viginta articuli de natura et mundo, (4) De lampade combinatoria et de specierum scrutinio (Wittenberg, 1587), (5) Animadversiones in lampadem lullianam ex codice Augustano nunc editae (1587. Part III contains (1) De lampade venatoria (Wittenberg, 1587), (2) De imaginum compositione (Frankfort, 1591), (3) Artificium perorandi (Frankfort, 1612—printed after Bruno's death).

Vol. III is in one part, which contains (1) Lampas triginta statuarum (1587?), (2) Libri physicorum Aristotelis explanati (1587?), (3) De magia et theses de magia (1589?), (4) De magia mathematica (1590), (5) De principiis rerum, elementis et causis (1590), (6) Medicina lulliana, (7) De vinculis in genere.

1888, Gottingen. Le opere di Giordano Bruno, ristampate da Paolo de Lagarde 2 vols. Vol. I contains (1) Candelaio (Paris, 1582), (2) La cena de le cener

(London, 1584), (3) De la causa, principio et uno (Venice, 1584), (4) De l'infinito, universo et mondi (Venice, 1584).
Vol. II contains (1) Spaccio de la bestia trionfante (Paris, 1584), (2) Caballa del cavallo pegaseo, con l'aggiunta dell'asino cillenico (Paris, 1585), (3) De gli eroici furori (Paris, 1585).

1907–1909, Bari. Opere italiane. Edited by G. Gentile and B. Spampanato. 3 vols.
Vol. I (Dialoghi metafisici, con note di G. Gentile) contains (1) Cena de le ceneri, (2) De la causa, principio e uno, (3) De l'infinito, universo e mondi.
Vol. II (Dialoghi morali, con note di G. Gentile) contains (1) Spaccio de la bestia trionfante, (2) Caballa del cavallo pegaseo, con l'aggiunta dell'asino cillenico, (3) De gli eroici furori.
Vol. III (Candelaio: commedia. Ediz. critica, con introduzione e note di V Spampanato) contains Candelaio.

PRIMARY SOURCES, OTHER PHILOSOPHERS

Aristotle. Works. Translated by W. D. Ross. Oxford, 1910–1928.

Avicebron. Fons vitae, ex Arabico in Latinum translatus ab. Iohanne Hispano et Dominico Gundissalino . . . ed. Clemens Baeumker, Aschendorff Monasterii, 1895. Beitrage zur Geschichte der Philosophie des Mittelalters, Vol. I, Parts II–IV.

Lucretius. De rerum natura. Translated by H. A. J. Munro. London, 1919.

Nicholas of Cusa. De docta ignorantia. Edited by E. Hoffman and R. Klibansky. Leipzig, 1932. Vol. I.

Plato. The Dialogues of Plato. Translated into English by Benjamin Jowett, Oxford.

Plotinus. Enneads. Translated from the Greek by Stephen McKenna. 5 vols. London, 1917–1930.

Saint Thomas Aquinas. Opera omnia. Edited by Vives. Paris, 1871–1880.

Saint Bonaventure. Opera omnia, ad claras aquas (Quaracchi), ex. typ. Colleggii S. Bonaventurae, 1882–1902.

SECONDARY SOURCES

For a complete listing of the works on Bruno prior to 1926, see V. Salvestrini, Bibliografia della opere di Giordano Bruno, Pisa, 1926; here only the "landmarks" are listed, and works published after 1926.

Aliotta, A. "Il problema dell'infinito," *La cultura filosofica*, V (1911), 205–232.

Atanassievitch, X. La doctrine metaphysique et geometrique de Bruno. Belgrade, 1923.

Bakewell, C. M. Source Book in Ancient Philosophy. New York, 1907.

Bartholmess, C. Jordano Bruno. 2 vols. Paris, 1847. Vol I on the life and times of Bruno; Vol. II on Bruno's works.

Berti, D. Vita di Giordano Bruno da Nola. Florence, 1868.

—— Documenti intorno a Giordano Bruno. Rome, 1880.

Boulting, W. Giordano Bruno. London, 1916.
Brunnhofer, H. Giordano Brunos Weltanschauung und Verhangnis aus den Quellen dargestellt. Leipzig, 1882.
Burckhardt, J. The Civilization of the Renaissance in Italy. New York, 1935.
Burtt, E. A. Metaphysical Foundations of Modern Physical Science. New York, 1927.
Carrière, M. Die philosophische Weltanschauung der Reformationszeit in ihren Beziehungen zur Gegenwart. Stuttgart, 1847; 2d edition, 1887.
Cassirer, E. Das Erkenntnisproblem. 2d edition. Berlin, 1911. Vol. I.
—— Individuum und Kosmos in der Philosophie der Renaissance. Berlin, 1927.
—— An Essay on Man. New Haven, 1944.
Charbonnel, J. R. L'ethique de Giordano Bruno et le deuxieme dialogue du spaccio. Paris, 1919.
—— La pensee italienne au XVI siecle et le courant libertin. Paris, 1919.
Clemens, F. J. Giordano Bruno und Nicolaus von Cusa. Bonn, 1847.
Corsano, A. Il pensiero di Giordano Bruno. Florence, 1940.
Dampier-Whetham, W. C. History of Science. New York, 1935.
Dilthey, W. Gesammelte Schriften. Leipzig-Berlin, 1914. Vol. II: Weltanschauung und Analyse des Menschen seit Renaissance und Reformation.
Duhem, P. Essai sur la notion de théorie physique de Plato à Galilée. Paris, 1908.
Fenu, E. Giordano Bruno. Morcelliana, 1938.
Frith, I. Life of Bruno. London, 1887.
Gentile, G. Il pensiero Italiano del rinascimento. Florence, 1940.
Gilson, E. La philosophie de S. Augustin. Paris, 1931.
—— The Philosophy of St. Bonaventure. Translated by Dom Illtyd Trethowan. New York, 1938.
—— The Spirit of Medieval Philosophy. New York, 1940.
Grassi, E. Giordano Bruno, Heroische Leidenschaften und Individuelles Leben. Bern, 1947.
Guzzo, A. (editor). Giordano Bruno, De la causa, principio e uno. Florence, 1933.
Hallam, H. Introduction to the Literature of Europe in the Fifteenth, Sixteenth, and Seventeenth Centuries. 4 vols. Paris, 1837–1839.
Höffding, H. A History of Modern Philosophy. Translated by B. E. Meyer. London, 1900.
Kristeller, P. O. The Philosophy of Marsilio Ficino. New York, 1943.
—— and J. H. Randall, Jr. "The Study of the Philosophies of the Renaissance," *Journal of the History of Ideas*, II (1941) 449–496.
Lovejoy, A. O. "The Dialectic of Bruno and Spinoza," *University of California Publications, Philosophy*, I (1904), 141–174.
—— The Great Chain of Being. Cambridge, Mass., 1936.
McIntyre, J. Giordano Bruno. London, 1903.
Mercati, A. Il sommario del processo di Giordano Bruno, con appendice di documenti sull'eresia e l'inquisizione a modena nel secolo XVI (*Studi e Testi*). Cittá del Vaticano, Biblioteca Apostolica Vaticana, 1942.

Mondolfo, R. "La filosofia di Giordano Bruno e la interpretazione di Felice Tocco," *La cultura filosofica*, V (1911) 450–482.
—— L'infinito nel pensiero dei greci. Florence, 1934.
—— "La filosofia di Giordano Bruno," *Minerva*, Vol. II (1944).
Namer, E. Les aspects de Dieu dans la philosophie de Giordano Bruno. Paris, 1926.
Olschki, L. Geschichte der neusprachlichen wissenschaftlichen Literatur. Halle, 1927. Vol. III, pp. 1–67.
—— "Giordano Bruno," *Deutsche Vierteljahrschrift fur Literaturwissenschaft und Geistesgeschichte*, II (1924), 1–78.
—— Giordano Bruno. Bari, 1927.
Randall, J. H., Jr. Making of the Modern Mind. Revised edition. New York, 1940.
Renda, A. (editor). Bruno, De la causa, principio e uno, introduzione e note a cura di Antonio Renda. Padua, 1941.
Ruggiero, Guido de. Giordano Bruno. Rome, 1913.
—— Storia della filosofia. Bari, 1940. Vol. III.
Saracista, M. La filosofia di Giordano Bruno nei suoi motivi plotiniani. Florence, 1935.
Sarauw, J. Der Einflus Plotins auf Giordano Brunos Degli Eroici Furori. Leipzig, 1916.
Schelling, F. Bruno, oder über das gottliche und natürliche Princip der Dinge. Berlin, 1802.
Spampanato, V. Vita di Giordano Bruno. Messina, 1921.
—— Documenti della vita di Giordano Bruno. Florence, 1933.
Symonds, J. A. The Renaissance in Italy. New York, 1935.
Taylor, H. O. Thought and Expression in the Sixteenth Century. 2 vols. New York, 1920.
—— The Medieval Mind. 4th edition. 2 vols. New York, 1925.
Thorndike, L. A History of Magic and Experimental Science. 6 vols. New York, 1923–1941.
Tocco, F. Le opere latine di Giordano Bruno esposte e confrontate con le italiane. Florence, 1889.
—— Le opere inedite di Giordano Bruno. Naples, 1891.
—— "Le fonti piu recenti della filosofia del Bruno," *Rendiconti della Reale Accademia dei Lincei*, I, Ser. V (1892), 503–581.
Toland, J. Collection of Several Pieces of Mr. John Toland, with Some Memories of His Life and Writings. London, 1726.
Troilo, E. La filosofia di Giordano Bruno. Turin, 1907.
Überweg, F. Geschichte der Philosophie. 12th edition. Berlin, 1924. Vol. III.
Windelband, H. Geschichte der Neueren Philosophie. Leipzig, 1919.
Wolf, A. History of Science, Technology, and Philosophy in the Sixteenth and Seventeenth Centuries. London, 1935.

INDEX

Bei Fragen zur Produktsicherheit wenden Sie sich bitte an:
If you have any questions regarding product safety,
please contact:

Walter de Gruyter GmbH
Genthiner Straße 13
10785 Berlin
productsafety@degruyterbrill.com